The experiences of working women are explored in this collection of essays by well-known feminist researchers. Tied together by the theme "place matters," the papers emphasize the social, cultural, economic, historical, and geographical contexts in which women work, focusing on the specific conditions under which their experiences emerge.

Topics include the transformation of the work force in nineteenth-century Montreal (Bettina Bradbury), feminization of skill in the British garment industry (Allison Kaye); the relationship between work and family for Japanese immigrant women in Canada (Audrey Kobayashi), experiences of women during a labour dispute in Ontario (Joy Parr), contemporary restructuring of the labour force in the United States (Susan Christopherson) and in an urban context in Montreal (Damaris Rose and Paul Villeneuve), the effect of gentrification on women's work roles (Liz Bondi), inequality in the work force (Sylvia Gold), and theoretical issues involved in understanding women in the contemporary city (Linda Peake). An introductory essay provides a review of current issues.

Women's organizations and policy planners as well as geographers, historians, and sociologists will find this book of great interest.

AUDREY KOBAYASHI is director of the Institute of Women's Studies, Queen's University.

Women, Work, and Place

Edited by

AUDREY KOBAYASHI

McGill-Queen's University Press
Montreal & Kingston • London • Buffalo

© McGill-Queen's University Press 1994
ISBN 0-7735-1225-X (cloth)
ISBN 0-7735-1242-X (paper)

Legal deposit 4th quarter 1994
Bibliothèque nationale du Québec

Printed in Canada on acid-free paper

Canadian Cataloguing in Publication Data

Kobayashi, Audrey, 1951–
  Women, Work, and Place
  Includes bibliographical references and index.
  ISBN 0-7735-1225-X (bound) –
  ISBN 0-7735-1242-X (pbk.)
  1. Women – Employment. 2. Women – Employment –
History. I. Kobayashi, Audrey Lynn, 1951–
  HD5053.W64 1994   331.4   C-94-900567-3

Part of chapter 4 has appeared previously in Joy Parr, *The Gender of Bread Winners: Women, Men and Change in Two Industrial Towns, 1880–1950:* (Toronto: University of Toronto Press 1990) and is published by permission of the author and University of Toronto Press.

Typeset in Sabon 10/12
by Caractéra production graphique, Quebec City

*The volume is dedicated to
the memory of John Bradbury*

Contents

Acknowledgments

Several years ago, John Bradbury popped into my office, as he often did in those days, to discuss the progress of a number of postgraduate students involved in research on women and work. In the course of the discussion, he suggested that we might come up with a project to focus some of the growing interest in this topic and to provide a forum in which we might explore some of the connections between geography and other disciplines. A few minutes later, John's idea had expanded to include plans for an international symposium and a subsequent book, and the project had a name: Women, Work, and Place. John immediately threw some of his considerable energy into the effort and began securing funding while we worked to organize the symposium. This book is the final result.

Within a few months John became ill, and his further participation was limited. Nonetheless, he attended the symposium, and provided a source of inspiration and always constructive criticism for all of us. In the ensuing weeks, he somehow found strength to continue our discussions, to read the drafts of the early papers, and to convey to me his vision of how this collection might finally take shape. The book is very much influenced and informed by his observations, therefore, and its dedication is more than a posthumous tribute: it is one more expression of how John Bradbury, as a scholar and as a warm, caring human being, influenced the world.

This collection has changed considerably since the time of its origins, and there are many people to thank for the final production. Papers have been added and updated and have benefited from the wisdom of a number of readers, including two anonymous referees whose insistent suggestions have resulted in genuine improvement. At the time of the symposium, students and colleagues at McGill braved one of the worst snow storms of the decade to ensure that the event would be a success. I thank especially Mireya Folch, Katherine Foster, Nina Laurie, Heidi Nast, Katie Pickles, Mark Rosenberg, Dermot Vibert, Jeanne Wolfe, and all of the participants – including those who contributed to a stupendous potluck supper, capped by André Levesque's tinned peaches and jello. Linda Peake provided advice and support and read many of the early drafts. Karen Smythe's skillful editing enhanced the text tremendously. At McGill-Queen's University Press, Joan McGilvray and Philip Cercone overcame many obstacles to keep production more or less on schedule. Helen Rowland, Karin Braidwood, and Phyllis Browne kept the project moving. Resources were provided by the Graduate Faculty of McGill University, the Canadian Studies program, and the Department of Geography. This book also owes a debt to all women who work, for it was inspired by their work and we have learned about the realities of everyday hard slogging from them. I have learned above all, to paraphrase Jane Nadel-Klein and Dona Davis (1988, 7), that women are not just passive weepers. The merits of this book are due to the contributors and the women from whom they have learned; I alone take credit for the errors and omissions.

Audrey Kobayashi
Montreal 1994

REFERENCE

Nadel-Klein, Jane and Dona Lee Davis. 1988. To work and to weep: women in fishing economies. Social and Economic Papers, no. 18. St John's, Nfld: Memorial University.

Introduction:
Placing Women and Work

AUDREY KOBAYASHI,
LINDA PEAKE, HAL BENENSON,
AND KATIE PICKLES

This collection presents interdisciplinary and international perspectives on women's lives, focusing on their work and the diverse places in which the complexities of gender relations are worked out. All of the contributors, a group that includes geographers, historians, and sociologists, are concerned with the ever changing circumstances of women's work and with the specific nature of these changes in different world regions, nation states, and urban areas.

In a similar collection, and in a somewhat similar vein, Reverby and Helly make the claim that history matters: "Before all women's historians disappear into the land of gender, language, binary oppositions, and representations, we need to remain ever mindful of the necessity of grounding our analysis in the material realities of class, race, sexuality, social structure, and politics" (1992, 16). Our perspective here is that this concept of *grounding* needs to be made much more explicit: in other words, that history without geography is an abstraction, and vice versa. By focusing on divergent milieux, therefore, these studies challenge unilinear stereotypes about women and work. Other major themes include the rethinking of conceptual categories such as the household economy and patriarchy; the persistence of material inequalities and women's attempts to effect changes; the power of cultural traditions in shaping "women's place"; and the significance of recent restructuring within production systems. Whereas some of the chapters focus specifically on aspects of women's employment, others examine the relations and combinations of waged and nonwaged

work, the public and private "spheres", and economic and sociocul-
tural patterns.

The methodologies represented here are diverse and reflect theoretical differences among the contributors as well as variation in their
academic disciplines and backgrounds. This variation, like gender,
is ultimately socially constructed; it also indicates an opening up of
feminism to a diverse range of enquiries, one which requires not only
that we abandon the uniformity encouraged by simplistic categorization of women's roles but that we shift both scale and method according to the dynamic social relations we hope to address. The individual
methodologies therefore range from extensive to intensive research and
quantitative to qualitative investigation, from national overviews
through local case studies to the intimacy of the household; they also
consider different time frames within several historical periods.

The papers are unified, nonetheless, in that it is gender relations,
and not women per se, that form the basis of our enquiries. The
reassessment and realignment of women's movements during the 1980s
has led to an academic shift from limited analyses of women's subordination and oppression in gender roles to studies of the ways in which
gender relations are socially constructed. Thus the papers are also
drawn together by their common focus on women as social *agents*.
This focus represents a political choice; as long as patriarchy remains
the dominant form of gender relations, we will need to address not
only its effects upon women but also the historical means by which it
has been constructed and the contemporary ways in which it is both
continued and combatted. Although the papers in this volume are not
in complete theoretical accord, then, there is a great deal of overlap;
and with the opening up of feminist analyses in general, there are
several theoretical implications for this book as a whole.

Studies of the work that permeates all aspects of women's lives have
contributed to an increasing recognition that the way we construct
theory is changing and that we need flexible analytical categories,
either because current ones do not fit the reality of women's daily lives,
or because they are too narrow to capture its complexity. There is now
greater diversity in women's lives than ever before, and therefore no
single theory of gender relations will explain the circumstances of all
women in all places or times. By addressing new areas, as well as
looking at old areas in new ways, our object of study ranges from
constraints upon women to the ways in which women create opportunities and structure their actions within those constraints, often
drawing upon past structures that include culturally embedded patriarchal relations. The reality of everyday life is a product of such limits
and opportunities; some strategies may allow women to challenge and

redefine their own circumstances, but such challenges are usually difficult. Although we may not have achieved what Lunneborg (1990) calls the "balance" that would mark a society in which there is no longer "women's work" and "men's work" and where power is not differentially structured along gender lines, nonetheless there is plenty of evidence – in these case studies and elsewhere in the feminist literature – that the framework is beginning to shift.

To understand this shift we need to understand that the political strategies endorsed and adopted by women's movements have also changed as definitions of "womanly" activities are continuously negotiated. Such redefinitions have led to a critical recognition of the limits of androcentricity as well as to reconsideration of older subjects of feminist analysis, such as notions of maternity. The relationship between production and reproduction, for example, has been reconsidered, and it is no longer possible to view these as separate spheres. The scope has expanded for understanding the tensions, satisfactions, and new social relations that have emerged in all areas of women's lives: there is increased attention given to women's positive experiences, in addition to the more painful and frustrating results of patriarchal gender relations, and to the ways in which the positive and painful experiences are mutually related. These are areas of analysis that traditionally have received little attention within the social sciences.

The necessity of putting the feminist critique into a multicultural context has also become clearer. In addition to studying the interaction between genders along economic/political/class lines, therefore, we here examine the link between gender and "race" or ethnicity where these features of social life converge in one place, or differ from place to place; such links not only etch the contours of cultural relations but form the basis of cultural politics in the struggle for equality. "Race" and ethnicity are socially constructed categories and not deterministic or "natural"; they are, to paraphrase Connell (1987, 111), enacted or conducted rather than simply expressed. This working definition allows us to recognize, indeed to emphasize, the active role of women and men as creators, rather than as mere bearers, of attributes such as sex, "race," or class, as well as their limited means and potential to control the ways in which such attributes define life conditions. All these more subtle points notwithstanding, however, racism and sexism combined form perhaps the single most widespread form of oppression in the world today; this fact is never far from our minds (see, for example, Amott and Matthaei 1991; Smith et al. 1988).

Emphasis on control and difference necessarily involves analyses of power, which suggest that patriarchal relations have come about not as simple relations of domination and subservience but as cultural conditions in which whole ways of life are bound up. There has been a shift, in recent feminist studies, from the acceptance of a reified notion of patriarchy as the key to understanding gender differences to an interest in the nature of the power by which patriarchal relations are structured. We join Joan Cocks (1989) in recognizing that the power to structure gender relations is not fixed in quantity or quality; that it is usually fractured along lines that do not conform to a single pattern; that it is contingent in time and place; and that it is almost never univocal. Moreover it is important to study power in the complex relationships of love and friendship and in the emotions – positive, negative, ambivalent – cultivated therein, as well as in institutional relationships between employer and employee, between citizen and state, and across the spectrum of public organizations that include, for example, the educational system and labour law.

It is equally important to recognize the ways in which these different dimensions of social life are mutually constructed, so that gender differences are empowered, not only along one dimension such as a gender division of labour, but in the infusion of that system with political, economic, traditional, and, not least, sexually erotic, values. Although some of the following chapters attempt to follow the linkages among these diverse realms of cultural life, our more limited aim here has been to tease out the ways in which the lines of power converge in the context of work itself.

The conditions under which women work in advanced, industrialized countries, and the work that they do, are changing rapidly. Since World War II women have entered the paid labour force in unprecedented numbers (Amott and Matthaei 1991; Armstrong and Armstrong 1990; Lewis 1992); the family is increasingly a site where new forms of gender relations, including equal parenting, are negotiated (Berger and Berger 1983; Chodorow 1978); and women are now more than ever aware of the political implications of their own actions and of their relationship to the political economy (Findlay 1987; Maroney and Luxton 1987). In addition women are beginning to challenge the traditional dichotomies that segregated their work according to the "paid" and the "unpaid" distinction and to question the assumption that unpaid work is less valuable (Waring 1988). At the same time older forms of labour, such as homeworking, are reemerging and affecting labour-market relations and industrial restructuring in ways that are appropriate to the contemporary economic scene (Lipsig-Mummé 1983; Peck 1992; see also Kaye, this volume); these changes reflect ethnic or racial divisions as much as those of class and gender. The restructuring of industry and paid labour, moreover, while difficult to capture or theorize at the macro level, has very specific local outcomes (Bagguley et al. 1990), and the "feminization of the work force" (Jenson, Hagen, and Reddy 1988) has vast implications for all areas of contemporary life; it is also seen, perhaps optimistically, as part of the larger sense in which the feminization of society may bring about fundamental changes in women's lives by shifting attitudes, values, and forms of human relationship such that they will take into account women's work experience (Segal 1987).[1] The following literature review examines some of the ways in which our conceptions of gender relations have altered during the last decade or so of rapid social change.

MYTHS ABOUT WOMEN'S WORK

A glance at the development of research in gender studies suggests the need for caution about overstating our present claims, because problematic assumptions about women's work are continually reappearing in new guises. Some of these assumptions in effect limit the challenge of studying women, work, and place by perpetuating certain myths about women's work, myths that have long been the foci of feminist critiques. Two prevailing myths especially have rendered women's work invisible: the myth of "*separate worlds*", or the idea that "public" occupational structures and arenas can be understood in isolation from so-called "private" or domestic family arrangements (Pleck 1976; Hall 1982; Davidoff 1979; Davidoff and Hall 1987; Mackenzie 1989); and the *masculine work norm*, which identifies production as a male domain and men's labour as the general standard for understanding women's work (Cockburn 1983 and 1985; Game and Pringle 1983).

What causes the old myths to reappear? Twentieth-century advanced industrialized societies have continually reproduced the division of sexual spheres in work and spatial organization and, because familiar, these divisions have seemed "natural," linking women to "private" families, to the residential milieu, and to domestic work (Harris 1981) and men to the "public" business district and to sites of production. The linkages are then reinforced through the actual creation of the built environment, as well as through place- and context-specific social practices. At the same time, women's movements have not easily challenged these divisions. Their mobilizations, even at their peak, have been partial, and they have encountered powerful resistance. They were more successful, for example, in gaining limited suffrage (for women over thirty) in Britain, in 1900–18 than they have been in securing equal pay in industry since then (Smith 1978); and efforts to legalize abortion in the United States in 1970–89 were better rewarded than those to obtain an Equal Rights Amendment – much less to acquire equal pay for jobs of comparable worth, to eliminate occupational segregation, to encourage the equal sharing of housework, or to refashion a built environment supportive of these changes (Greenberger 1980; Hayden 1981; Pollack Petchesky 1984; Michelson 1985; Fox and Fox 1986; Tienda and Ortiz 1987; Lupri and Mills 1987).

These practical blockages have had theoretical consequences. Political obstacles and the resilience of gender hierarchies have hampered the outcome expected by intellectual analysts and have thwarted their intentions of fully comprehending the social and spatial bases of gender divisions. Obviously, then, we need to study not only the empirical contexts of women and work but also the intellectual contexts in which our understanding is circumscribed as much by our political experience as by our analytical acuity.

Three examples of fundamental research on women and work illustrate this point. Each was important in its time as an overview of change in women's work, but each, while breaking new ground, at the same time reproduced problematic assumptions in new forms. They are: Ivy Pinchbeck's *Women Workers in the Industrial Revolution* (1930; rpt., 1969), the first scholarly treatment of the topic; Alva Myrdal and Viola Klein's *Women's Two Roles* (1956; rev., 1968), the first cross-national study of married women's entry into employment after 1945; and Louise A. Tilly and Joan W. Scott's *Women, Work and Family* (1978; rev., 1987), which updated both Pinchbeck's and Myrdal and Klein's analyses. Each of these studies set out to bring to light women's economic contribution to society and to challenge one or both of the prevailing myths about separate worlds and masculine work norms.[2] As a prelude to the studies presented in this book, it is worthwhile to review these three critical studies both to establish an intellectual heritage and to illustrate the ways in which social and analytical myths are interconnected.

Ivy Pinchbeck sought to demonstrate women's centrality to British industrialization during the period from 1750 to 1850, stressing wage-earning women's accomplishments. Pinchbeck documented the range of conditions confronting women in agriculture, textile and metal trades, homework industries, coal mining, and small businesses. She also recognized, but did not take into account in her analysis, geographic variation in the structure of work environments.

Pinchbeck's perspective reflected both feminist influences and the state of the British women's movement in the 1920s. While gains were made in the century under study, especially for educated and wage-earning women, women's accomplishments met with resistance. The gains included the campaign for the "endowment of motherhood," which demanded that women's work in the home be recognized as economically vital (Lewis 1980); the admittance of women to the legal and other professions under the Sexual Disqualification (Removal) Act of 1919 (Strachey 1988); the granting of degrees to women at Oxford University in 1920 (Leonardi 1989); and the unprecedented organization of wage-earning women during the First World War (Stevenson 1984). But women's accomplishments had met resistance. After the First World War a masculine economic backlash had blocked women's further entry into nontraditional employment (Smith 1984) and had generalized the principle of the "marriage bar," which divided married women (who now had to resign from jobholding) from the unmarried. The latter became the typical career women (Martindale 1939).

The challenges faced by women's mobilizations provided the background for Pinchbeck's analysis, but her own ability to challenge the norm of separate sexual spheres remained limited. Pinchbeck rationalized gender hierarchy in her conclusion, stating that married women's removal from paid work and their return to economic dependence on husbands had led to the valuation of women's work in the home. Thus, for Pinchbeck, wives' domesticity represented a gain: "the industrial revolution marked a real advance since it led to the assumption that men's wages should be paid on a family basis, and paved the way for the more modern conception that in the rearing of children and in home making, the married woman makes an adequate economic contribution" (1969, 312–13). By believing that this valuation was genuine and widely accepted, Pinchbeck engaged in wishful thinking. Her argument, moreover, obscured the extent to which the norm of the male family wage (which prescribed men's exclusive support of families) denoted female economic dependence.

Pinchbeck's assessment of the impact of industrialization on unmarried women was also problematic. She argued that this group had achieved an individualistic independence: "In the case of the single working woman, the most striking effect of the industrial revolution was her distinct gain in social and economic independence ... Under the new regime every woman received her own earnings ... The significance of this change was at once seen in the new sense of freedom which prompted so many young women to retain control of their wages, and even to leave home at an early age in order to become their own 'mistresses'" (1969, 313). But Pinchbeck's analysis reflected more accurately the anxieties of middle-class observers (preoccupied with the moral effects of female wage-earning and autonomy) than the real situations of young working-class women (Eisenstein 1983). The latter most commonly remained part of parental households and conceived new aspirations in terms of a collective work group rather than individualized achievements. Pinchbeck's analysis imputed a male, individualist model to single women workers that ignored gender differences in wage earning. It also accepted the marriage bar as an unalterably given fact: men's political opposition allowed single women to have job aspirations that married women could not. This rigid separation of single and married women also overlooked the obvious point that, for Pinchbeck's period of study, single working women often later married and thus that their tenure of supposed independence was, in fact, short-lived.

Myrdal and Klein (1968) attempted to challenge the prevailing myths but also reproduced old assumptions in new guises. They produced the first cross-national study to recognize and seek to explain the fundamental shift in married women's labour force involvement after 1945. They challenged the notion that employment issues were men's problems, and reconceptualized the changing relationships between work and family life. Their insights into the structural causes of women's changing employment – women's decreasing fertility, increased longevity – and their arguments for state provision of collective services (nurseries, canteens, community housing) reflected social changes of the post-Second World War era. In Britain, for example, these changes included the dismantling of the marriage bar in government employment, partly in response to women's demands (Parris 1973); the state's increasing dependence on married women's labour in wartime and post-war export manufacture; and the mobilization of Labour Party and trade union women for the maintenance of day nurseries used by employed mothers (Summerfield 1984). On the negative side, however, the "welfare state" consensus demobilized the women's movement, and treated married women workers as a secondary work force that had to adjust to pronatalist state policy and husbands' expectations that wives would take the vast share of responsibility for housework and child rearing (Riley 1984).

In this setting the critical elements in *Women's Two Roles* were submerged in a psychological validation of separate spheres that identified married women with families and procreation (Beechey 1987). The core of Myrdal and Klein's analysis was the thesis that an inner conflict of roles based on different normative expectations was experienced uniquely by women: "The characteristic feminine dilemma of to-day is usually summarized under the heading 'Career and Family.' The struggle for the right to work is no longer directed against external obstacles; no longer is there the same hostile public opinion to overcome with which our grandmothers had to contend, nor is there a lack of opportunities for women. To-day the conflict has become 'internalized' and continues as a psychological problem ... the pull in two directions goes on practically throughout a woman's life" (1968, 136). By treating the readjustment of work and family relationships on this essentially psychological level, Myrdal and Klein obscured the basis of women's position in the unequal division of labour in the home and in segregated employment. The view of role conflict as an exclusively feminine dilemma leaves unanswered why *men's* work and family relationships were unproblematic. Underlying this asymmetry is Myrdal and Klein's naturalistic identification of women with biology. In their words, women's "lives were more intricately linked with the existence of the family and the continuation of the race [than were men's lives]" (1968, xv). Thus Myrdal and Klein recreate a division of spheres; woman's psychological juggling act as manager of two conflicting roles (family and career) is contrasted with men's unquestioned association with employment. Subsequent and recent role conflict studies have reiterated Myrdal and Klein's questionable assumptions (Barnett and Baruch 1985; Cooke and Rousseau 1984; Elman and Gilbert 1984; Epstein 1974, 475–8; Gray 1983; Hall 1972; Harrison and Minor 1978; Van Meter and Agronow 1982; for critical assessments, see Connell 1985; Ferree 1990, 867–8).

Tilly and Scott (1987) later attempted to revise these individualistic accounts of women's experience through a reexamination of the demographic sources of women's employment. They sought to clarify women's response to economic change and, especially, to reveal the interconnectedness of work and family activities. In addition they developed a useful typology of urban milieux, such as textile centres with many jobs for women, mining communities with few opportunities, and commercial cities with home-based crafts. Underlying their approach was a critical response to the "modernization" perspective, which claims that Western societies promote female emancipation on the basis of individualistic values. In contrast Tilly and Scott reflected the broader current of 1970s women's political activism in the United States, emphasizing the persistence of social inequality, material constraints, and the lack of individualism in women's lives. This context helps account for their critical insights but, at the same time, their reliance on a static notion of the family economy falsified women's experience and recreated separate spheres through a naturalistic identification of women with the family.

The crux of their analysis is two-fold. First, they denied that industrialization had changed women's work: "The aggregate view indicates that industrialization *did not change* the type of work women did, nor did it greatly increase the amount of time that women in the aggregate devoted to productive work for market exchange. Indeed over the course of the nineteenth century women's work force participation varied very little ... [T]hroughout the nineteenth century in England and France the majority of working women performed jobs with low levels of skill and low productivity similar to those that had characterized women's work for centuries" (1987, 77; emphasis added). Secondly, they asserted that the family context of women's decision making had remained constant over the period of their analysis (1700–1960s): "in all periods different kinds of work for women represented family economic strategies of adaptation to changed conditions; family strategies, in other words, were *continuous* phenomena in relation to women's ... work" (1987, 6; emphasis added). But some of the evidence they present calls these generalizations into question. For example they describe young women engaged in textile manufacture who bargained with their parents "on terms ... of equality" (120) and who "were more autonomous in their choice of husbands than daughters had been in the past" (122).

Tilly and Scott also failed to explore both the extent to which the factory system had created radically new social relationships for women (involving, for example, supervision by foremen and contacts with many other women at work) and that to which mechanization had altered both their labour processes (for example, in work as power-loom weavers) and their economic standing (as female workers who were paid at the same base rates as males) (Rubinstein 1986; Roberts 1988). Furthermore their argument that women textile workers constituted a minority of the nineteenth-century female work force (Tilly and Scott 1987, Table 4-2), and that "the majority of working women remained in household settings" (77), overlooked the exemplary importance in England of textile workers, who constituted a reference point in discussions of all women workers' situations.[3] Their view also obscured new developments – such as urbanization and modern phenomena associated with the rise of a wealthy middle class and the entry of young working-class women into domestic service (McBride 1976) – that were *not* explicable simply as "the *persistence* of a characteristic form of women's employment ... from the preindustrial to the industrial period" (6; emphasis added).[4]

Tilly and Scott's continuity thesis also overlooked the impact of industrialism on relations between male and female workers, who were no longer part of common household units of production (as in the putting-out system) but, rather, frequently confronted one another as potential job competitors. This situation often provoked restrictive measures against women on the part of male trade unionists (Taylor 1983; Prothero 1979; Hewitt 1975; Boston 1980; Lewenhak 1977). By evading discussion of relations between male and female workers, Tilly and Scott masked this development. Furthermore their "continuity" argument, which treated women as the agents of "family strategies," became clearly untenable for the period from 1945 to the 1960s. In this era of decreased fertility, a supposed diminution in the heavy burden of housework and rising rate of divorce altered the bases on which adult women made work decisions and heightened the possibilities for gender conflict and for women's assertion of collective, work-related demands (Kiernan 1988; Rimmer and Popay 1982; Meehan 1985; Borzeix and Maruani 1988; Phillips 1988).

Yet Tilly and Scott maintained the myth that a static, unitary "family economy" dictated women's decisions in the present as in the past. Their argument for this period fails not only to recognize diversity in family arrangements – for example, it does not mention divorced women or single mothers supporting households – but also identifies women in a timeless way with the family and the life cycle, while leaving men's involvement out of the picture. Their "one-gender history" thus reproduces an implicit separation of spheres of family-identified women and invisible men and obscures the fact of gender inequality as a determinant of women's work patterns. Later research employing their family economy perspective, however, has attempted to redress some of these limitations (see Hill 1989, 24–68).

CONNECTIONS AND INTERCONNECTIONS

The above critique was undertaken to show that, taken as a whole, Pinchbeck's, Myrdal and Klein's, and Tilly and Scott's studies have been crucial to our understanding of women's place in the world of work, but that the authors encountered difficulty in fully comprehending the social forces underlying women's position in the division of labour. In their focus on women's paid employment these researchers achieved some success in defusing the myth of the masculine work norm, but they perhaps unwittingly assumed the existence of gender-divided separate spheres. In failing to link the two domains or to develop an encompassing analytic framework, they have left this challenge open to others.

We return, then, to our original question: What causes the old myths to reappear in contemporary research? We may also ask, given the conceptual difficulties and analytical inadequacies of the separate spheres myth, why feminists continue to engage in the spheres debate with such zeal; or why it seems, as Carole Pateman has observed, that "[t]he dichotomy between the private and the public is central to almost two centuries of feminist writing and political struggle; it is, ultimately, what the feminist movement is about" (1989, 118). If this is so then we need to ask whether this statement represents a factual description of where most feminist efforts have been expended, which it seems to do, and whether it also stands as a normative statement of the most pressing issue facing feminist researchers today, in which case there is a need for a much more extended and critical form of analysis. In this vein we must confront the separate spheres challenge on at least two fronts. First, historically, the separation of public and private was a material condition that came about (although never simply, completely, or without resistance) through the realization of patriarchal social norms based on the concept that women's place is in the home. Its origin dates at least to the time of classical Greece, when the separation of the public world of Man, politics, and freedom from the private world of women, domesticity, and necessity was justified according to women's putative inferiority and man's (also putative) superior penchant for civilization (Cox 1987, 7; Elshtain 1984, 4). From that time through the "Enlightenment" and Victorian periods, much male effort was spent to create and value a patriarchal society that could thwart the "disorder" that would surely result should woman be allowed to exceed her place (Pateman 1989). Rousseau's notion of the "sexual contract," applied as a means of maintaining proper separation of the sexes, has been reformulated and reinstated over and over again so that such separation has become valued as a fundamental and "natural" aspect of human life (Pateman 1988), so much so that basic social institutions such as the family have, in many places, drawn their archetypes and formulated their strategies for survival from the separate spheres model (see examples in Hudson and Lee 1990). As most of the papers in this collection illustrate, most women and men today continue to order their lives with role separation and according to a masculine work norm. Even individuals who depart from this life pattern are often only grudgingly admitted as participants to social processes defined by the male experience (Hudson and Lee 1990, 2).

But there is a second, critical, reading of this domination that, as suggested above, can compel us to go beyond the patriarchal ideology of separate spheres and to explore alternatives to its replication even within the ideology of feminism. As Kerber (1988, 39) notes, "One day we will understand the idea of separate spheres as primarily a trope, employed by people in the past to characterize power relations for which they had no other words and that they could not acknowledge because they could not name, and by historians in our own times as they groped for a device that might dispel the confusion of anecdote and impose narrative and analytical order on the anarchy of inherited evidence, the better to comprehend the world in which we live." Recent attempts to dispel that confusion have been partially successful; but it is by no means clear that we have succeeded in escaping the bonds of the separate spheres mythology. "The feminist total critique of the liberal opposition of private and public still awaits its philosopher," writes Pateman (1989, 136). There seems to be a growing consensus, nonetheless, that "the metaphor of separate spheres has been stretched too far. It explains too little" (Kerber 1989, 566), and we are now in the process of developing alternative analytic models.

One form of resistance to the intellectuel tyranny of hierarchical spheres has been the growth of "women's culture" as a field. Beginning in the 1970s some feminists, influenced by the work of Myrdal and Klein but determined to cast woman's role in a different light, questioned the designation of the private sphere as inferior, and attempted to show that there was an alternative to the study of men's public world that had previously been the only space considered worthy of study. The uncovering of women's experiences soon developed into a new perspective, largely undertaken by historians who proposed that women's past "cannot be presumed to mirror the portion of male reality that most historians have chosen to highlight" (Strong-Boag and Fellman 1986, 3). Women's concepts of time and place have differed considerably from those of their male counterparts (see, for example, Cott 1977; Prentice et al. 1988), and interpretations of what takes place in shared spaces have also differed, sometimes profoundly (Conrad et al. 1988). Therefore understanding women's experiences in the past requires redefinition of the concepts of time and space to make them appropriate to women's experiences and to places of kinship, home, and community – places to which women have given meaning and where they have exerted influence (Ball 1975; Campbell 1990).

As well as studying women's work, the women's culture approach has viewed the celebration of rituals and traditions, such as assisting with childbirth and exchanging recipes, as "a source of strength and identity that afforded supportive sisterly relations" (Cott 1977, 12). The biological realities of pregnancy, childbirth, nursing, and menopause have bound women together through history (Smith Rosenberg 1975; see also Jensen 1986). Furthermore work on women's sexuality that does not define it against male-generated norms has suggested that a uniquely female social realm exists and provides women with "support, intimacy and ritual" (Smith Rosenberg 1975, 28; see also Seidman 1990; 1991). Such bonds are not static but have changed in response to different conditions at different times and places (Griffin Cohen 1988; Jensen 1986).

The uncovering of women's history has been one of the most important projects of feminist scholars. The works cited here, and others, have played an important role in presenting and valuing women's lives and work. Faced with the challenge that women have often been silenced in the public records, such works have also utilized alternative sources such as quilts, samplers, oral histories, and diaries (Cott 1991; Harris and Phillips 1984), thereby developing new and valuable methodological tools. Nonetheless the women's culture approach is problematic in a number of ways. Its one-gender perspective ignores or denies the role of men, thus eluding both the issue of how people are gendered *in relation* to one another and that of *how* patriarchal relations have structured and dominated the domestic sphere. This romantic approach certainly endorses the notion of separate spheres, then; though it places greater value upon femininity, it often ignores the lives of those for whom such a positive and romantic image may not be possible or even desirable. The sources, such as diaries and handicrafts, are often those produced in middle-class settings, where women's work can easily be portrayed as women's success. There is no doubt that many women have aspired to such settings (see Kobayashi, this volume), but they are unattainable, or attained only at great cost, for many women of colour, for women from working-class families, and for many single and elderly women. This approach also casts maternity and women's sexuality in normative terms. Such norms, upheld as much by middle-class White women as by men, often have led only to "bitter choices" for those women to whom work means "to earn a living for themselves and their families" and to work a double day in the process (Rosen 1987, 2).

Other feminists, in contrast, have advocated either the rejection or the eradication of the private sphere. Some liberal and radical feminists have a surprising amount in common on this issue: the former argue for social means to enable women to enter the superior public world of men in traditionally male-occupied roles (e.g., Cox 1987); some of the latter (e.g., Firestone 1970; Millett 1970) claim that patriarchy develops from the family, the place of which must be radically restructured to free women from the bonds of biology. Clearly these two positions are diametrically opposed in the way that the family is viewed (liberal feminists want access to the privileged world of men by protecting and freeing women to pursue both home and career; radical feminists reject such traditional values altogether), but both reformulate the separate spheres myth, using it as a foundational analytical tool. Socialist feminists, in contrast, argue for a transformation of the overall position of women in society by way of changing women's status both at home and in the labour market. Such changes require the breakdown of the current social and spatial division of labour, as well as the creation of new forms of built environments (Women and Geography Study Group 1984).

Still others have shown that the creation of spheres need not occur on patriarchal lines, nor by celebrating women's culture, but by structuring gender *relations* so as to create egalitarian outcomes. Anderson (1987), in a study of seventeenth-century Huron women, argues for egalitarian, "non-incursive" gender relations through which "gender specific tasks [are] mediated through complementary sets of kin relations" (126). To admit a range of variability is also to admit that there are no simple explanations (such as, for example, the status of women as mothers or women's integration into the capitalist system) for social/ spatial situations. Rather there is need to understand "the ways in which women and men are both divided from each other and then reunited into functioning economic, political, and emotional structures" (Anderson 1987, 126). One condition of this understanding is the recognition that gender divisions of labour occur in more than two spheres. Kayn and Gozemba (1992), for example, challenge the notion of separate spheres on the grounds both of place and of sexuality in their study of the working-class lesbian bar culture. Gannagé's (1987) case study of a Canadian garment factory adds the union to the home and workplace as sites where the gender division of labour is constructed, and cross cut, by other forms of division, such as ethnicity. Gannagé's study shows, however, that ethnic division is not, as it is often portrayed, simply a mechanism for splitting the labour market, however important that process may be (Bonacich 1972; Boswell and Jorjani 1988); it also opens new avenues for challenging gender ideologies, as demonstrated in her example wherein union strategies learned in Southern Italy were applied to a Canadian context.

Throughout the 1980s feminist studies of women and work have shifted the emphasis from general overviews to focused case studies, thus going some way to overcoming the separate spheres myth by exposing the conditions under which the separation of home and workplace – associated with functions of reproduction and production – has become a social norm. Examinations of conditions as they exist in both spheres have revealed the web of processes that define conditions of life, albeit differentially, for many women (Mackenzie and Wajcman 1985; Pollert 1981; Luxton 1980; Tsurumi 1990; Westwood 1985). Although such case studies examine various circumstances, their overarching theme concerns the interconnectivity of the two spheres, and the many ways in which the normative ideal of separation is either rejected or unattainable. Having identified connections as crucial, however, feminist scholars then faced the challenge of analyzing women's changing position in a variety of contexts and as part of a single process. Recent work has addressed the tendency toward bifurcation along the axes of domestic environment and paid employment by siting the intersection of the two spheres within a single context (Lowe and Gregson 1989; Tsurumi 1990). Many studies of women's work continue to fall clearly into one of the two categories, but we now consider these to be interconnected rather than separate spheres.

Perhaps most important of all, the connectivity of spheres as theorized in the gender relations approach has been borne out by the general incorporation of the notion of patriarchy into our thinking, and by a growing understanding of the many and varied circumstances under which patriarchy manifests itself. Sylvia Walby (1990), for example, distinguishes between two forms of patriarchy, the public and the private or, in Ursel's (1988) terms, the familial and the social. This analytical distinction provides a refined sense of those conditions under which patriarchy develops, and allows our emphasis to shift to the *sites* in which gender relations are played out rather than focusing, as was the case in earlier work, upon the roles appropriate or expected within particular spheres. With this new focus, *context* becomes of crucial importance.

This brings us to the spatial organization of women's work. Incipient notions of spatiality, such as that found in Tilly and Scott's (1987) comparative urban typology, have been strengthened by the increasingly sophisticated understanding of variations in patriarchy according to site and context. Throughout the 1980s feminist geographers have added a further element to the study of women's work and, especially, of the sources of separations between domestic and market work environments. But while their spatial analyses have opened new perspectives, they have also fallen prey to the pitfalls discussed above. An example illustrates this problem. Historical geographers have identified the differentiation of nineteenth-century urban space – which separated workers' slum dwellings, middle-class suburbs, factories, and central commercial streets – as a pivotal aspect of industrialization. One of its effects was the expression of domesticity in spatial terms, as an attribute of residential areas (Women and Geography Study Group 1984; Davidoff and Hall 1987), although this expression had a range of ramifications according to social group. Nonetheless the varied emergence of domesticity as separate from large-scale workplaces isolated women's domestic work (Hakim 1985; Deacon 1985), and set in place a trajectory of separation, the logical outcome of which was the extreme isolation found in the many contemporary suburbs where women's activities and opportunities are severely curtailed by their spatial confinement (Everitt 1976). As a result attention has focused on the definition as opposed to the interaction of places, or on how women literally move from place to place and the logistics of their movement (Michelson 1988; Palm 1979).

Given the continued reality of spatial disjunction in women's lives, then, it is important not to erase theoretically either the separation of women's spheres or the sexual division of labour, for these are real separations created under specific material conditions of life; such separations have resulted in significant adjustments in people's lives and have affected gender relations profoundly (Hanson and Pratt 1991; Meissner et al. 1988; Pratt and Hanson 1991a, b). Furthermore they have generally created greater hardship for women, who not only must juggle the demands of the home and workplace but must do so under structural, and spatial, conditions that have traditionally facilitated men's working conditions at the expense of women's (Dyck 1989, 1990). Examination of the evidence for changing gender relations and working conditions under restructured forms of capitalism has emphasized recently how resilient are the practices of patriarchy. Doreen Massey's pathbreaking work, *Spatial Divisions of Labor*, provides a brilliant discussion of the ways in which social/class relations are organized and gendered to produce geographically differentiated patterns of employment, and of employment change (1984, 12–13). The *re*production of inequality draws upon, *inter alia*, women's "lack of previous experience of wage relations, the masculinity and sexism of the local culture," which were "preconditions established under the dominance of the previous spatial divisions of labour" (300). These observations have led to widespread recognition of the reproduction of patriarchy in new forms (Walby 1989) and, among geographers especially, of the need to connect the specificity of the local with the structural conditions of global change (Bowlby 1990; Stubbs and Wheelock 1990; see also Bagguley et al. 1990; Murgatroyd et al. 1985). By doing so, however, we may recognize that culture can transcend location as something that is shared, although its specific attributes may differ, among people in a number of locations. This point is made dramatically by a collection of essays on women in fishing communities around the world who share not only a common vocation but similar familial and personal experiences (Nadel-Klein and Davis 1988).

Against this enthusiastic embrace of locality studies as a means of connecting the local and the global, Linda McDowell has cautioned that "[c]urrent definitions of patriarchy themselves, at least as applied to economic restructuring, have proved an inadequate theorization of contemporary changes, disguising the ways in which new divisions in the labour market and in the home are opening up among and between women and men. The continued assumption of those feminist theorists who rely on the notion of patriarchy that class and gender divisions may be separately theorized (and by Marxists who see class interests

as predominant) disguises the ways in which they are mutually constituted. If this is understood a more dynamic view of the different ways
in which women's and men's interests are divided or united at different
times is possible" (1991, 416). McDowell's critique points out what
many of the most sophisticated feminist analyses have missed: that an
understanding of the changing conditions of women's work is impossible without a full-scale understanding of the overall conditions under
which capital production is organized. Furthermore, as Judith Stacey
(1992) has shown, changing forms of production also influence the
form of feminism itself. Her statement that geographical expressions
of changing work need to be understood at a number of scales and
according to the variety of experiences of individuals and groups at
particular places, "spatially *and* socially differentiated" and "divided
by age, by race and ethnic origins and by their social and cultural
experiences" (417; emphasis added), should be a clarion call; but it is
disappointing to realize that there is as yet so little of such empirical
work that allows context and connection to be explored in particular
places. Part of the challenge of this book, therefore, was to imagine
how such studies could be done.

The emphasis on interaction among places and the problems thereby
encountered leads to questions about how public policy and the legal
system have (in the past) reflected the ideology of women's domesticity
in order to encourage a separation of spheres and how they might (in
the future) be used to overcome gender divisions. Recent work on
public policy emphasizes the need for provision and improvement of
social programmes that affect child care (Mackenzie and Truelove
1993), housing (Klodawsky and Spector 1988), and reproduction
(Zigler and Frank 1988). The fallacy of the spheres debate is illustrated
dramatically upon consideration of these issues, traditionally regarded
as private but made matters of state interest in the political sphere
where "[w]omen as a class do not share power" (Spallone 1988, 156).
The politics of reproduction have engendered huge debates on the
need for either protective or special legislation, and on the hidden
disadvantages of various methods that enable women to overcome the
production/reproduction dichotomy. Heitlinger (1987) reveals the
value of a comparative perspective to show, not only that outcomes
vary under different political and economic regimes, but that the
definition of what is deemed "productive" or "reproductive" varies
according to context-specific ideological systems. Mackenzie (1988)
and Roberts (1991), among others, claim that what is required is no
less than a rebuilding of the physical environment in order to shift the
basis for gender constitution, thereby creating what Wekerle et al.
(1980) have aptly termed "new space for women."[5]

It is at this point that the notion of *place* comes into its own. An understanding of the sites at which patriarchal practices are enacted requires not only that context be treated as a background, but also that the siting and situating of such practices recognize the constitutive role of the place itself as inseparable from social outcomes. This dialectical understanding of the thoroughly material conditions of social life provides a theoretical means, not only of going beyond some of the more difficult dichotomies involved in comprehending social relations (agency/structure, ideal/material, etc., as well as separate spheres), but also the methodological means of understanding empirical conditions more fully. Furthermore the recognition that conditions differ from place to place provides us with a more powerful means of understanding differences among women, thus forestalling the development of feminist stereotypes and recognizing that a wide variety of processes, virtually always overlapping and interlocking, serve to define difference.

SHEDDING THE MYTHS (AGAIN)

If the Enlightenment entrenched the ideology of domesticity and separate spheres, it also reinforced a tendency to think in dualistic terms. We are still recovering from these effects. The compulsion to separate, therefore, is not only a dominant ideology that separates men and women and creates the conditions that allow patriarchal domination; it is also the ideological lens through which analytical categories have been distorted into divergent streams. The problems of what Ray Pahl (1988, 7) calls our "present confusions about the meaning of work," therefore, can be attributed both to the difficulty of analyzing shifting social conditions and to the difficulty of transcending theoretical blockages.

To exemplify the logical problems encountered as a result of dualistic categories, Chris Middleton (1988, 24) advances three conflicting premises about the relationship between the often separated categories of gender divisions and capital accumulation:

1 Gender divisions in the labour market are a by-product of processes of capital accumulation and/or class struggle.
2 Segmented labour markets are created by processes of capital accumulation and/or class struggle. Gender divisions become superimposed on the market for reasons which lie *beyond the scope of the model*.
3 Gender divisions are not derivative of capital accumulation and/or class struggle. They are independently responsible for structuring labour markets according to 'patriarchal' principles (although that term may not always be used).

A similar logic has been used by feminist geographers, who have had a difficult time overcoming the legacy of dualism. One of the most significant developments in feminist geographical theory centred on the "*Antipode* debate," which traded critical readings on the salience of patriarchy and capitalism (Foord and Gregson 1986; McDowell 1986; Knopp and Lauria 1987; Gier and Walton 1987; Johnson 1987; Gregson and Foord 1987) to show that feminist geographers, too, have had a difficult time overcoming the legacy of dualism in their theoretical debates.

To view capitalism and patriarchy as separate and/or opposing categories, however, is to preclude a set of possibilities whereby:

1 There is no *necessary* relationship between patriarchy and capitalism; one or the other may play a more significant role depending upon circumstances.
2 Patriarchy and capitalism are dialectically interconnected, so that it is impossible to define one without reference to the other.
3 Patriarchy and capitalism change form (albeit slowly, given their coercive power and durability) over time, or are worked out differently in different contexts.
4 Other structuring processes gain greater explanatory power under certain circumstances. Such processes may be affiliative in nature, such as kinship or cultural practice (neither defined monolithically), or they may be exclusionary and subordinating, such as racism or fascism.

To thus explode the sense of dualism and to suggest a variability of explanatory frameworks in this way, however, does more than suggest a commitment to contextual analysis and contingent explanation; it also confronts the problems of essentialism. The arguments about spheres, after all, were arguments about what categories could be considered essential, not about whether we could transcend the concept of categories. We need to shift the focus of the debate, therefore, to ask why we feel compelled to identify essentialist categories, and how feminist ideology re-inscribes old dualisms.

To state a long and complicated discussion in abbreviated terms, we need to look at the dominant Western modes of thought, at the heritage, stretching from Hellenic times to the Enlightenment to the present, by which patriarchy, dualistic thinking, and an obsession with primordial causes have been invariably associated. That association is characterized by the theoretical attempt to place diverse "woman" into limited analytical categories, which is an intellectual reductionism that is fundamentally connected to the moral reductionism that confines women's actions and asserts an ideologically defined "woman's place." But to transcend categorical analysis we need to confront the processes through which the attribution of essential "truth" is naturalized and given intellectual credibility.

There is still room for analytical categories, as the above discussion has shown, both because they reflect empirical conditions and because we seem to need to work through them to achieve theoretical progress. And, ironically, without categories we also run the political risk of making our subjects invisible by denying their commonality. But the challenge is twofold: we must utilize categories that are meaningful to and that will address the problems of contemporary women; and we need to establish a dialectical relationship between what we read as material, historical categories and our emerging analytical system. In the process we might hope to avoid creating new myths about women's work that will impede understanding in the future. Of course we will likely fail in this regard, but we can hope at least to be critically aware of our failures.

This critical path will lead us to a reconceptualization of myth. Heretofore we have been concerned with debunking the myths that have obscured our analytical vision of the conditions of women's work. If we hope, however, that by clearing away the cloud of mythology we shall expose some fundamental or essential truth, we shall be disappointed. The critical agenda calls instead for a form of deconstruction of myths about women's work that will allow us to engage with those ideological terms through which the myths were constructed and thereby will bring us to understand something about myth-making as social discourse. For myth-making is, to paraphrase Peter Fitzpatrick (1992, 1), the reconciliation of contradictory existences. The big gender myths – separate spheres, the male work norm, and the male sense of superiority upon which patriarchy is based – all have provided means of naturalizing the contradictory existences of women and men. It is their natural quality – as a set of stories that not only seem to make intuitive sense but set didactic and moral terms – upon which ideological practices build.

Those practices, which marshal power to exclude, appropriate, homogenize, or essentialize women (Spelman 1988; Spivak 1988, 118), constitute our culture in the broadest sense of the term; they shape the way(s) we do things and our system(s) of linking our pasts and our future aspirations. Rather than treat culture as either an essential or a determinant body of practices or artifacts (as it is treated in the women's culture approaches discussed above), culture should be taken as the constant production and reproduction of a system of communication, with all that communication implies: power, structure, connection. Culture – the filter of myth, the bastion of ideology and unreflective practice, and the vehicle for group constitution – is a continuous discourse and involves the construction, constitution, and expression of commonality. It gains analytic power if seen in Spivak's (1988) terms of "cultural politics." Culture needs to be taken very seriously, then, as the process that sets the conditions under which women find themselves working.

But we have come full circle here, to the by now self-evident point that culture occurs in place; it has a geography and spatial structure; it is not an abstract rationale for understanding symbolic edifices once removed from their physical expression but, rather, a system of reference for particular women and men who do not simply have culture, but *constitute* it. Our project as feminists, to use Bondi and Domosh's terms, is to "explore and contextualize the literal and metaphorical spatiality of contemporary discourse" (1992, 210), that is, to interrogate the texts of women's lives that are produced through cultural discourse, to bring to life their experiences in such a way that they are cultured, gendered, spatialized – in short, contextualized.

WOMEN, WORK, AND PLACE

The essays in this collection cannot make a collective claim to having dispelled all the myths of women's work or to have uncovered all the new myths recently created. Certainly they can make only the most modest of claims to have unravelled some of the difficult theoretical issues identified in this introduction. Nonetheless they all offer a perspective on the problem of *placing* women and work, thus enriching the contextual understanding that we have claimed is crucial to progress in the study of women and work. Each case study adds its own approach and contribution to this overarching theme.

The collection opens with a chapter by Linda Peake, who directly addresses the issue of separate spheres in her discussion of those ways in which urban analysts have constructed social theory. She advances her own perspective, focusing on the inadequacies of Manuel Castells's work for understanding recent changes concerning women and work in Britain.

This theoretical piece is followed by several chapters that illustrate the historical conditions under which women have worked. Bettina Bradbury observes local variations in employment opportunities and the impact of new technologies on nineteenth-century Montreal homes and industries employing women. She argues that technological changes, accompanied by a reorganization of work and scale of production, did not substantially alter women's lives either in terms of paid employment or domestic work. Audrey Kobayashi also considers links between home and work in her documentation of the interlocking webs of patriarchal relations that structured Issei women's experience in Canada, while simultaneously trying to understand how their subordination was experienced as both oppressive and a source of happiness.

In a further examination of the links between home and work, Joy Parr looks at the interplay between family values of respectability and mill grievances in the shaping of womanly militancy amongst women knitting workers in Paris, Ontario, in 1949. She argues that labour militarism was selectively reinterpreted and depoliticized so as to reproduce traditional gender relations and stereotypes. Both Parr and Kobayashi emphasize the points that not only labour relations but human relations are central to our understanding of the experience of work, and that that experience is filtered through an emotional lens that, as an important element of social experience, should not be ignored. That lens, moreover, is filtered by the cultural practices of community participation. Finally Sylvia Gold provides a political analysis of Canadian women's positions as employees and as entrepreneurs, linking women's roots in the feminist movement to their situation in contemporary Canada where, despite significant changes in social, economic, legal, and productive contexts, inequalities and patriarchal practices persist.

The balance of chapters has a contemporary focus, situating conditions of change in four countries at a number of spatial scales. Alison Kaye's study of the garment industry in the East End of London, England, focuses on the social construction of skill in the context of shifting gender and racial divisions of labour. Her argument is that the definition of women's skills in paid employment is influenced by the fact that their training is hidden away in the home, thereby causing the work to appear *un*skilled, and that their skill is racialized and thus subordinated within the overall labour market. This study provides an example of how the sphere of women's work is constructed and normalized. Also at the urban level, Damaris Rose and Paul Villeneuve examine the relationship between job location and skill levels for employed Montreal women. With an emphasis on occupational restructuring in different parts of the metropolis, they explore material on contemporary suburban/city-centre differences in employment opportunities.

Susan Christopherson uses another spatial scale, exploring national and regional changes in women's employment in the United States. She focuses specifically on the contribution of households in nonmetropolitan areas of Arizona in her attempt to evaluate the impact of economic restructuring. The final paper, by Liz Bondi, examines at a theoretical level gender roles and relations in the context of women's involvement in urban gentrification. She focuses on the ways in which occupational restructuring and changing gender relations lead to a broader urban restructuring; she also brings us back to our original points about our framework: that it is necessary to integrate analytical notions of process with context in order to provide a fuller understanding of women's lives, and that the integration of spheres involves more than a simple recognition of the public/private dichotomy.

Taken together these studies show the remarkable variety and contingency that characterize women's life and work in different places. To identify this variation, however, is not to suggest that the places involved, like the infamous separate spheres, are isolated or unique. Each specific place is a complex mixture of circumstances linked in various ways to larger processes. Our objective, therefore, has been to recognize the ways in which abstract social processes, often international in scope, are always *placed*, lived, and given specific meaning by specific people in landscapes that are constructed in relation to the world at large. We need to look therefore, at such powerful developments as capitalism or industrialization and their attendant ideologies, which affect women's work profoundly and are "not bounded by time and geography" (Franzoi 1987, 153), to show that each place has much in common with other places (see also Gabaccia 1987; Meyer 1987). This point about shared characteristics of place is increasingly important in today's world of internationalized economic practices, standardization of production methods, and time/space compression (Harvey 1989).

Such commonality will not be understood, however, by using patriarchal history as a standard against which to measure patriarchal effects; as Boxer and Quataert (1987, 95) point out, the "standard markers of change may have affected women's lives little, if at all, for women rarely participated in political or military struggles." Throughout the collection, therefore, efforts have been made to look at large processes that change the material conditions under which women work at specific times and places, as well as at the ideological changes that set new parameters for guiding human action and outlining a public sense of what is right, good, and natural. But these are seldom straightforward changes, and, although we may identify dominant trends and tendencies that influence (sometimes overwhelmingly) the circumstances of women's lives – what we call history – we may also see the extent to which those circumstances are resisted, contested, and negotiated (Rose 1990). The ground upon which the negotiation of gender relations occurs emerges as an important focus of this collection.

The notion of context as a basis for addressing this much neglected aspect of the constitution of work in different places, therefore, is the central theme of this book. These contributions have attempted to move away from generalized, top-down analyses and to focus upon the interaction of processes in the cultural, political, and economic settings of different places on a number of scales, from the local to the international. We recognize that the range of countries presented here – Britain, Canada, and the U.S., all advanced industrial countries of the North – limits the nature of our enquiry. Yet, while our purpose is not primarily to provide comparative analyses, these places contain enough diversity for discussion of our major concerns. On the other hand research on the South, for example, would open up some very different areas of concern, and we are not unaware of the continuing need to expand our horizons (see Mohanty 1991).

Still, this book is about particular women in particular places. We hope that it presents a collective effort and shows, whatever other issues are tabled upon the ongoing agenda of feminist studies, that place matters, and that a woman's place is what she makes it.

NOTES

1 The study of the feminization of society requires not only the substudy of femininities as models – although such limited studies might be useful (see, for example, Grossman and Chester 1990) – but also the substudy of masculinities both as a basis for patriarchy and as being subject to change under new forms of gender relations. This approach is adopted in more recent work by Segal (1990). See also Jackson (1991), who draws upon the concept of cultural politics to explore the construction of gender relations and to show that, despite the political challenges of feminism, patriarchal structures are remarkably resilient.

2 The critique of these assumptions is not new but, rather, goes back to Olive Schreiner's *Women and Labour* (1911), which had been influenced by the early twentieth-century suffrage movement.

3 For an example of the centrality of textile workers to social debate, see Margaret Hewitt (1975).

4 Domestic servants constituted 40 per cent of the female labour force in Britain in 1851 and 22.5 per cent in France in 1866 (*Women, Work and Family* Table 4-2).

5 Despite this work, however, Pratt and Hanson have pointed out that "arrangements within the household have received little explicit attention" (1991b, 55).

REFERENCES

Amott, Teresa and Julie Matthaei. 1991. *Race, Gender, and Work*. Montreal and New York: Black Rose Books.

Anderson, Karen. 1987. "A gendered world: women, men, and the political economy of the seventeenth-century Huron." In *Feminism and Political Economy*, Heather Jon Maroney and Meg Luxton, eds, 121–38. Toronto, New York, London, Sydney, Auckland: Methuen.

Armstrong, Pat and Hugh Armstrong. 1990. *Theorizing Women's Work*. Toronto: Garamond Press.

Bagguley, P., J. Mark-Lawson, D. Shapiro, J. Urry, S. Walby, and A. Warde. 1990. *Restructuring: Place, Class and Gender*. London: Sage.

Ball, R. 1975. "A perfect farmer's wife: women in nineteenth-century rural Ontario." *Canada: An Historical Magazine* 3. 2: 2–21.

Barnett, R.C. and G.K. Baruch. 1985. "Women's involvement in multiple roles and psychological distress." *Journal of Personality and Social Psychology* 49: 135–45.

Beechey, Veronica. 1987. *Unequal Work*. London: Verso.

Berger, B. and P.L. Berger. 1983. *The War over the Family: Capturing the Middle Ground*. London: Hutchinson.

Bonacich, Edna. 1972. "A theory of ethnic antagonism: the split labor market." *American Sociological Review* 37. 5: 547–9.

Bondi, L. and M. Domosh. 1992. "Other figures in other places: on feminism, postmodernism and geography." *Environment and Planning D: Society and Space* 10: 199–213.

Borzeix, Anni and Margaret Maruani. 1988. "When a strike comes marching home." In *Feminization of the Labour Force: Paradoxes and Promises*, Jane Jenson et al., eds, 245–59. New York: Oxford University Press.

Boston, Sarah. 1980. *Women Workers and Trade Unions*. London: Davis-Poynter.

Boswell, Terry and David Jorjani. 1988. "Uneven development and the origins of split labor market discrimination: a comparison of black, Chinese, and Mexican immigrant minorities in the United States." In *Racism, Sexism, and the World-System*, Joan Smith et al., eds, 169–86. New York, Westport, CT, and London: Greenwood Press.

Bowlby, S.R. 1990. "Technical change and the gender division of employment: the new information technology industries in Britain." *Geoforum* 21. 1: 67–84.

Boxer, Marilyn and Jean Quataert, eds, 1987. *Connecting Spheres: Women in the Western World, 1500 to the Present*. New York: Oxford University Press.

Campbell, Gail G. 1990. "Canadian women's history: a view from Atlantic Canada." *Acadiensis* 20. 1: 184–99.

Chodorow, Nancy. 1978. *The Reproduction of Mothering*. Berkeley: University of California Press.

Cockburn, Cynthia. 1983. *Brothers: Male Dominance and Technological Change*. London: Pluto Press.

– 1985. *Machinery of Dominance*. London: Pluto Press.

Cocks, Joan. 1989. *The Oppositional Imagination: Feminism, Critique and Political Theory*. London and New York: Routledge.

Connell, R.W. 1985. "Theorizing gender." *Sociology* 19: 260–72.

– 1987. *Gender and Power*. London: Polity.

Conrad, M., T. Laidlaw, and D. Smyth. 1988. *No Place Like Home*. Halifax: Formac Publishing.

Cooke, R.A. and D.M. Rousseau. 1984. "Stress and strain from family roles and work-role expectations." *Journal of Applied Psychology* 69: 252–60.

Cott, Nancy. 1977. *The Bonds of Womanhood: "Woman's Sphere" in New England, 1780–1835*. New Haven: Yale University Press.

Cox, S., ed. 1987. *Public and Private Worlds: Women in Contemporary New Zealand Society*. Wellington: Allen and Unwin.

Davidoff, Leonore. 1979. "The separation of work and home? Landladies and lodgers in nineteenth- and twentieth-century England." In *Fit Work for Women*, Sandra Burman, ed, 64–97. London: Croom Helm.

Davidoff, Leonore and Catherine Hall. 1987. *Family Fortunes: Men and Women of the English Middle Class, 1750–1850*. Chicago: University of Chicago Press.

Deacon, Delsey. 1985. "Political Arithmetic." *Signs* 11. 1 (Autumn): 27–47.

Dyck, I. 1989. "Integrating home and wage workplace: women's daily lives in a Canadian suburb." *The Canadian Geographer* 33. 4: 329–41.

– 1990. "Space, time and renegotiating motherhood: an exploration of the domestic workplace." *Environment and Planning D: Society and Space* 8. 4: 459–84.

Eisenstein, Sarah. 1983. *Give us Bread, But Give us Roses*. London: Routledge and Kegan Paul.

Elman, M.R. and L.A. Gilbert. 1984. "Coping strategies for role conflict in married professional women with children." *Family Relations* 33: 317–27.

Elshtain, J.B. 1981. *Public Man, Private Woman: Women in Social and Political Thought*. Princeton: Princeton University Press.

Epstein, Cynthia Ruchs. 1974. "Reconciliation of women's roles." In *The Family: Structure and Functions*, Rose Laub Coser, ed, 473–89. New York: St. Martin's Press.

Everitt, J.C. 1976. "Community and propinquity in a city." *Annals of the Association of American Geographers* 66. 1: 106–16.

Ferree, Myra Marx. 1990. "Beyond separate spheres: feminism and family research." *Journal of Marriage and the Family* 52: 866–84.

Findlay, Sue. 1987. "Facing the state: the politics of the women's movement reconsidered." In *Feminism and Political Economy: Women's Work, Women's Struggles*, Heather Jon Maroney and Meg Luxton, eds, 31–50. Toronto, New York, London, Sydney and Auckland: Methuen.

Firestone, Shulamith. 1970. *The Dialectic of Sex: The Case for Feminist Revolution*. New York: Morrow.

Fitzpatrick, Peter. 1992. *The Mythology of Modern Law*. London: Routledge.

Foord, J. and N. Gregson. 1986. "Patriarchy: towards a reconceptualisation." *Antipode* 18. 2: 186–211.

Fox, Bonnie J. and John Fox. 1986. "Women in the labour market, 1931–81: exclusion and competition." *The Canadian Review of Sociology and Anthropology* 23. 1: 1–21.

Franzoi, Barbara. 1987. "' … With the wolf always at the door … ': women's work in domestic industry in Britain and Germany." In *Connecting Spheres: Women in the Western World, 1500 to the Present*, Marilyn Boxer and Jean Quataert, eds, 146–55. New York, Oxford: Oxford University Press.

Gabaccia, Donna. 1987. "In the shadows of the periphery: Italian women in the nineteenth century." In *Connecting Spheres: Women in the Western World, 1500 to the Present*, Marilyn Boxer and Jean Quataert, eds, 166–76. New York: Oxford University Press.

Game, Ann and Rosemary Pringle. 1983. *Gender at Work*. London: George Allen and Unwin.

Gannagé, Charlene. 1987. "A world of difference: the case of women workers in a Canadian garment factory." In *Feminism and Political Economy*, Heather Jon Maroney and Meg Luxton, eds, 139–65. Toronto, New York, London, Sydney, Auckland: Methuen.

Gier, J. and J. Walton. 1987. "Some problems with reconceptualising patriarchy." *Antipode* 19. 1: 54–8.

Gray, J.D. 1983. "The married professional woman: an examination of her role conflicts and coping strategies." *Psychology of Women Quarterly* 7: 235–43.

Greenberger, Marcia. 1980. "Effectiveness of federal laws in the United States." In *Equal Employment Policy for Women*, Ronnie Steinberger Ratner, ed, 108–28. Philadelphia: Temple University Press.

Gregson, N. and J. Foord. 1987. "Patriarchy: comments on critiques." *Antipode* 19. 3: 371–5.

Griffin Cohen, M. 1988. *Women's Work, Markets and Economic Development in Nineteenth-Century Ontario*. Toronto: University of Toronto Press.

Grossman, Hildreth Y. and Nia Lane Chester, eds, 1990. *The Experience and Meaning of Work in Women's Lives*. Hillsdale, NJ: Lawrence Erlbaum Associates.

Hakim, C. 1985. "Social monitors." In *Essays on the History of British Sociological Research*, Martin Bulmer, ed, 39–51. Cambridge: Cambridge University Press.

Hall, Catherine. 1982. "The butcher, the baker, the candlestick maker: the shop and the family in the industrial revolution." In *The Changing Experience of Women*, Elizabeth Whitelegg et al., eds, 2–16. Oxford: Martin Robertson.

Hall, D.T. 1972. "A model of coping with role conflict: the role behaviour of college educated women." *Administrative Science Quarterly* 17: 471–86.

Hanson, S. and G. Pratt. 1991. "Job search and the occupational segregation of women." *Annals of the Association of American Geographers* 81. 2: 229–53.

Harris, Olivia. 1981. "Households as natural units." In *Of Marriage and the Market: Women's Subordination in International Perspective*, K. Young, C. Wolkowitz, and R. McCullagh, eds, London: CSE Books.

Harris, R.C. and E. Phillips, eds. 1984. *Letters from Windermere, 1912–1914*. Vancouver: University of British Columbia Press.

Harrison, A. and J. Minor. 1978. "Inter-role conflict, coping strategies and satisfaction among Black working wives." *Journal of Marriage and the Family* 40. 4: 799–805.

Harvey, David. 1989. *The Condition of Postmodernity*. Oxford and Cambridge, Mass.: Basil Blackwell.

Hayden, Dolores. 1981. "What would a non-sexist city look like? Speculations on housing, urban design and human work." In *Women and the American City*, Catharine R. Stimpson et al., eds, 167–84. Chicago: The University of Chicago Press.

Heitlinger, Alena. 1987. "Maternity leaves, protective legislation, and sex equality: Eastern European and Canadian perspectives." In *Feminism and Political Economy*, Heather Jon Maroney and Meg Luxton, eds, 247–62. Toronto, New York, London, Sydney, Auckland: Methuen.

Helly, Dorothy O. and Susan M. Reverby, eds. *Gendered Domains: Rethinking Public and Private in Women's History*. Ithaca and London: Cornell University Press.

Hewitt, Margaret. 1975. *Wives and Mothers in Victorian Industry*. Westport, CT: Greenwood Press.

Hill, Bridget. 1989. *Women, Work and Sexual Politics in Eighteenth-Century England*. Oxford: Basil Blackwell.

Hudson, Pat and W.R. Lee. 1990. *Women's Work and the Family Economy in Historical Perspective*. Manchester and New York: Manchester University Press.

Jackson, Peter. 1991. "The cultural politics of masculinity: towards a social geography." *The Institute of British Geographers Transactions*, n.s., 16. 2: 199–213.

Jensen, T.M. 1986. *Loosening the Bonds*. New Haven: Yale University Press.

Jenson, Jane, Elizabeth Hagen, and Ceallaigh Reddy, eds. 1988. *Feminization of the Labour Force*. Oxford: Polity Press.

Johnson, L. 1987. "(Un)realist perspectives: patriarchy and feminist challenges in geography." *Antipode* 19. 2: 210–15.

Kayn, Janet and Patricia A. Gozemba. 1992. "In and around the lighthouse: working-class lesbian bar culture in the 1950s and 1960s." In *Gendered Domains: Rethinking Public and Private in Women's History*, Dorothy O. Helly and Susan M. Reverby, eds, 90–108. Ithaca and London: Cornell University Press.

Kerber, Linda J. 1988. "Separate spheres, female worlds, woman's place: the rhetoric of women's history." *Journal of American History* 75. 1: 9–39.

– 1989. Untitled section of "Beyond roles, beyond spheres: thinking about gender in the Early Republic," by Linda J. Kerber, Nancy F. Cott, Robert Gross, Lynn Hunt, Caroll Smith-Rosenberg, and Christine M. Stansell. *William and Mary Quarterly* 46. 3 (July): 565–85.

Kiernan, Kathleen A. 1988. "The British family: contemporary trends and issues." *Journal of Family Issues* 9. 3: 298–316.

Klodawsky, Fran and Aron Spector. 1988. "New families, new housing needs, new urban environment: the case of single-parent families." In *Life Spaces: Gender, Household, Employment*, Caroline Andrew and Beth Moore Milroy, eds, 141–58. Vancouver: University of British Columbia Press.

Knopp, L. and M. Lauria. 1987. "Gender relations as a particular form of social relations." *Antipode* 19. 1: 48–53.

Leonardi, Susan J. 1989. *Dangerous by Degrees*. New Brunswick, NJ: Rutgers University Press.

Lewenhak, Sheila. 1977. *Women and Trade Unions*. New York: St. Martin's Press.

Lewis, Jane. 1980. *The Politics of Motherhood*. London: Croom Helm.

– 1992. *Women in Britain Since 1945*. Oxford and Cambridge, Mass.: Blackwell.

Lipsig-Mummé, C. 1983. "The renaissance of homeworking in developed countries." *Relations Industrielles* 38: 545–67.

Lowe, M. and N. Gregson. 1989. "Nannies, cooks, cleaners and au pairs: new issues for feminist geography." *Area* 21: 415–17.

Lunneborg, Patricia W. 1990. *Women Changing Work*. New York, Westport, CT, London: Bergin & Garvey.

Lupri, Eugene and Donald L. Mills. 1987. "The household division of labour in young dual earners couples: the case of Canada." *International Review of Sociology* 2:

Luxton, Meg. 1980. *More than a Labour of Love: Three Generations of Women's Work in the Home*. Toronto: The Women's Press.

Mackenzie, Donald and Judy Wajcman, eds. 1985. *The Social Shaping of Technology: How the Refrigerator Got its Way*. Milton Keynes: Open University Press.

Mackenzie, Suzanne. 1988. "Building women, building cities: toward gender sensitive theory in the environmental disciplines." In *Life Spaces: Gender, Household, Employment*, Caroline Andrew and Beth Moore Milroy, eds, 13–30. Vancouver: University of British Columbia Press.

– 1989. "Restructuring the relations of work and life: women as environmental actors, feminism as geographic analysis." In *Remaking Human Geography*, Audrey Kobayashi and Suzanne Mackenzie, eds, 40–61. Boston, London, Sydney and Wellington: Unwin Hyman.

Mackenzie, S. and M. Truelove. 1993. "Changing access to public and private services: non-family childcare." In *The Changing Social Geography of Canadian Cities*, Larry S. Bourne and David F. Ley, eds, 326–42. Montreal, Kingston, London and Buffalo: McGill-Queen's University Press.

Maroney, Heather Jon. 1987. "Feminism at work." In *Feminism and Political Economy*, Heather Jon Maroney and Meg Luxton, eds, 85–107. Toronto, New York, London, Sydney, Auckland: Methuen.

Maroney, Heather Jon and Meg Luxton. 1987. "From feminism and political economy to feminist political economy." In *Feminism and Political Economy: Women's Work, Women's Struggles*, Heather Jon Maroney and Meg Luxton, eds, 5–28. Toronto, New York, London, Sydney, Auckland: Methuen.

Martindale, Hilda. 1939. *Woman Servants of the State*. London: George Allen and Unwin.

Massey, Doreen. 1984. *Spatial Divisions of Labor: Social Structures and the Geography of Production*. New York: Methuen.

– 1991. "Flexible sexism." *Environment and Planning D: Society and Space* 9: 31–57.

McBride, Theresa. 1976. *The Domestic Revolution*. London: Croom Helm.

McDowell, L. 1986. "Beyond patriarchy: a class-based explanation of women's subordination." *Antipode* 18. 3: 311–21.

– 1991. "Life without father and Ford: the new gender order of post-Fordism." *Transactions of the Institute of British Geographers*, n.s., 16. 4: 400–19.

Meehan, Elizabeth M. 1985. *Women's Rights at Work: Campaigns and Policy in Britain and the United States*. London: Macmillan.

Meissner, Martin, Elizabeth W. Humphreys, Scott M. Meis, and William J. Scheu. 1988. "No exit for wives: sexual division of labour and the cumulation of household demands in Canada." In *On Work: Historical, Comparative and Theoretical Approaches*, R.E. Pahl, ed, 476–95. Oxford and New York: Basil Blackwell.

Meyer, Sibylle. 1987. "The tiresome work of conspicuous leisure: on the domestic duties of the wives of civil servants in the German Empire (1871–1918)." In *Connecting Spheres. Women in the Western World, 1500 to the Present*, Marilyn Boxer and Jean Quataert, eds, 156–65. New York and Oxford: Oxford University Press.

Michelson, William. 1985. *From Sun to Sun: Daily Obligations and Community Structure in the Lives of Employed Women and Their Families*. Totowa, NJ: Rowna and Allanheld.

– 1988. "Divergent convergence: the daily routines of employed spouses as a public affairs agenda." In *Life Spaces: Gender, Household, Employment*, Caroline Andrew and Beth Moore Milroy, eds, 81–101. Vancouver: University of British Columbia Press.

Middleton, Chris. 1988. "The familiar fate of the famulæ: gender divisions in the history of wage labour." In *On Work: Historical, Comparative and Theoretical Approaches*, R.E. Pahl, ed, 21–47. Oxford and New York: Basil Blackwell.

Millett, Kate. 1970. *Sexual Politics*. Garden City, NY: Doubleday.

Mohanty, C.T. 1991. *Third World Women and the Politics of Feminism*. Bloomington: Indiana University Press.

Murgatroyd, L., M. Savage, R. Shapiro, J. Urry, S. Walby, and A. Warde. 1985. *Localities, Class and Gender*. London: Pion.

Myrdal, Alva and Viola Klein. 1968. Rev. ed. *Women's Two Roles*. London: Routledge and Kegan Paul. Orig. ed., 1956.

Nadel-Klein, Jane and Dona Lee Davis. 1988. To work and to weep: women in fishing economies. Social and Economic Papers, no. 18. St. John's, Nfld: Memorial University.

O'Brien, Mary. 1989. *Reproducing the World: Essays in Feminist Theory*. Boulder: Westview Press.

Pahl, R.E. 1988. "Historical aspects of work, employment, unemployment and the sexual division of labour." In *On Work: Historical, Comparative and Theoretical Approaches*, R.E. Pahl, ed, 7–20. Oxford and New York: Basil Blackwell.

Palm, Risa. 1979. "The daily activities of women." In *Women and the Social Costs of Economic Development: Two Colorado Case Studies*, Elizabeth Moen, Elise Boulding, Jane Lillydahl, and Risa Palm, eds, 99–118. Boulder, Colorado: Westview Press.

Parris, Henry. 1973. *Staff Relations in the Civil Service*. London: George Allen and Unwin.

Pateman, Carole. 1988. *The Sexual Contract*. Stanford: Stanford University Press.

– 1989. *The Disorder of Women*. Cambridge: Polity Press.

Peck, J.A. 1992. "'Invisible threads': homeworking, labour-market relations, and industrial restructuring in the Australian clothing trade." *Environment and Planning D: Society and Space* 10. 6: 671–89.

Phillips, Paul and Erin Phillips. 1983. *Women and Work: Inequality in the Labour Market*. Toronto: James Lorimer.

Phillips, Roderisk. 1988. *Putting Asunder*. Cambridge: Cambridge University Press.

Pinchbeck, Ivy. 1969. Rpt. *Women Workers in the Industrial Revolution.* London: Frank Cass. Orig. ed., 1930. London: G. Routledge and Sons Ltd.

Pleck, Elizabeth. 1976. "Two worlds in one: work and family." *Journal of Social History* 10. 2: 178–95.

Pollack Petchesky, Rosalind. 1984. *Abortion and Woman's Choice: The State, Sexuality and Reproductive Freedom.* Boston: Northeastern University Press.

Pollert, Anna. 1981. *Girls, Wives, Factory Lives.* London: MacMillan.

Pratt, G. and S. Hanson. 1991a. "Time, space and the occupational segregation of women: a critique of human capital theory." *Geoforum* 22. 2: 149–58.

– 1991b. "On the links between home and work: family strategies in a buoyant labour market." *International Journal of Urban and Regional Research* 15. 1: 55–74.

Prentice, Alison, Paula Bourne, Gail Cuthbert Brandt, Beth Light, Wendy Mitchinson, and Naomi Black. 1988. *Canadian Women: A History.* Toronto: Harcourt, Brace, Jovanovich.

Prothero, Iowerth. 1979. *Artisans and Politics in Early Nineteenth-Century London.* Folkestone, Kent: Dawson.

Rathbone, Eleanore. 1984. Rpt. *The Disinherited Family.* Bristol: Falling Wall Press. Orig. ed., 1927. London: G. Allen and Unwin.

Riley, Denise. 1984. *War in the Nursery.* London: Virago.

Rimmer, Lesley and Jennie Popay. 1982. Employment trends and the family. Study Commission on the Family, occasional paper 10. London.

Roberts, Elizabeth. 1988. *Women's Work, 1840–1940.* London: Macmillan.

Roberts, M. 1991. *Living in a Man-Made World.* Andover, Hants: Routledge, Chapman and Hall.

Rose, G. 1990. "The struggle for political democracy: emancipation, gender and geography." *Environment and Planning D: Society and Space* 18: 395–408.

Rosen, Ellen Israel. 1987. *Bitter Choices: Blue-Collar Women in and out of Work.* Chicago and London: The University of Chicago Press.

Rubinstein, David. 1986. *Before the Suffragettes.* New York: St. Martin's Press.

Schreiner, Olive. 1911. *Woman and Labour.* London: T. Fisher Unwin.

Segal, Lynne. 1987. *Is the Future Female?* London: Virago.

Seidman, Steven. 1990. "The power of desire and the danger of pleasure: Victorian sexuality reconsidered." *International Journal of Social History* 24. 1: 47–67.

– 1991. *Romantic Longings: Love in America, 1830–1980.* New York: Routledge.

Smith, Harold. 1978. "The issue of 'equal pay for equal work' in Great Britain, 1914–1919." *Societas* 8. 1 (Winter): 39–51.

– 1984. "Sex vs. class: British feminists and the labour movement, 1919–1929." *The Historical* 47. 1: 19–37.

Smith, Joan, Jane Collins, Terance K. Hopkins, and Akbar Muhammad, eds,
1988. *Racism, Sexism, and the World-System*. New York, Westport, CT, and
London: Greenwood Press.

Smith-Rosenberg, Carrol. 1975. "The female world of love and ritual: relations
between women in nineteenth-century America." *Signs* 1. 1: 1–29.

Spallone, Patricia. 1988. *Beyond Conception: The New Politics of Reproduction*. Granby, Mass.: Bergin & Garvey.

Spelman, Elizabeth V. 1988. *Inessential Woman: Problems of Exclusion in
Feminist Thought*. London: The Women's Press.

Spivak, Gayatri Chakravorty. 1988. *In Other Worlds: Essays in Cultural
Politics*. New York and London: Routledge.

Stacey, Judith. 1992. "Sexism by a subtler name?: postindustrial conditions
and postfeminist consciousness in the Silicon Valley." In *Gendered Domains:
Rethinking Public and Private in Women's History*, Dorothy O. Helly and
Susan M. Reverby, eds, 322–38. Ithaca and London: Cornell University
Press.

Stevenson, John. 1984. *British Society, 1914–1945*. New York: Penguin.

Strachey, Ray. 1988. Rpt. *The Cause*. London: Virago. Orig. ed., 1928.

Strong-Boag, Veronica and A. Clair Fellman, eds. 1986. *Rethinking Canada:
The Promise of Women's History*. Toronto: Copp Clark Pitman.

Stubbs, Cherrie and Jane Wheelock. 1990. *A Woman's Work in the Changing
Local Economy*. Aldershot, Hants and Brookfield, Mass.: Avebury.

Summerfield, Penny. 1984. *Women Workers in the Second World War*. London:
Croom Helm.

Taylor, Barbara. 1983. *Eve and the New Jerusalem*. New York: Pantheon.

Tienda, Marta and Vilma Ortiz. 1987. "Intraindustry occupational recompo-
sition and gender inequality in earnings." In *Ingredients for Women's
Employment Policy*, Christine Bose and Glenna Spitze, eds, 23–51. Albany,
NY: SUNY Press.

Tilly, Louise A. and Joan W. Scott. 1987. Rev. ed. *Women, Work and Family*.
New York: Methuen. Orig. ed., 1978. New York: Holt, Rinehart and
Winston.

Tsurumi, E. Patricia. 1990. *Factory Girls: Women in the Thread Mills of Meiji
Japan*. New Jersey: Princeton University Press.

Ursel, Jane. 1988. "The state and the maintenance of patriarchy: a case study
of family, labour and welfare legislation in Canada." In *Gender and Society:
Creating a Canadian Women's Sociology*, Arlene Tigar McLaren, ed, 108–
45. Toronto: Copp Clark Pitman.

Van Meter, M.J.S. and S.J. Agronow. 1982. "The stress of multiple roles: the
case for role strain among married college women." *Family Relations* 31:
131–8.

Walby, Sylvia. 1989. "Flexibility and the sexual division of labour." In *The Transformation of Work?*, S. Wood, ed, 127–40. London: Unwin Hyman.
– 1990. *Theorizing Patriarchy.* London: Blackwell.
Waring, M. 1988. *Counting for Nothing: What Men Value and What Women are Worth.* Wellington: Allen and Unwin.
Wekerle, Gerda R. 1988. "Canadian women's housing cooperatives: case studies in physical and social innovation." In *Life Spaces: Gender, Household, Employment*, Caroline Andrew and Beth Moore Milroy, eds, 102–40. Vancouver: University of British Columbia Press.
Wekerle, Gerda R., Rebecca Peterson, and David Morley, eds. 1980. *New Space for Women.* Boulder: Westview.
Westwood, Sallie. 1985. *All Day, Every Day.* Urbana and Chicago: University of Illinois Press.
Women and Geography Study Group of the Institute of British Geographers (IBG). 1984. *Geography and Gender.* London: Hutchinson.
Zigler, Edward F. and Meryl Frank. 1988. *The Parental Leave Crisis: Toward a National Policy.* New Haven and London: Yale University Press.

1 Engendering Change: Women's Work and the Development of Urban-Social Theory

LINDA PEAKE

To what extent do theories of urban-social change assist our under-standing of the ways in which various social relations, such as those differentiated by class and gender, are experienced and attributed meaning in people's everyday lives – especially in their places of work? Any consideration of this question requires a review and critique of the evolution of urban-social theory, from which we can then examine the categories used to measure urban-social change. The first part of this essay explores the initial three stages in the development of urban-social theory, as illustrated by the work of Manuel Castells. There are two reasons for this choice of theorist.[1] First Castells's work is repre-sentative of much research in urban-social studies, and his shifting focus has been of central importance in reorienting the object of study in the field and in determining the agenda for research on urban-social change. Secondly his work clearly exemplifies the ways at each of these three stages in which theories of urban-social change have approached the issue of transformation in socioeconomic structures and reciprocal changes in social relations. In the remainder of my essay, I argue that a fourth stage of theory, offered by feminist perspective, helps to transcend certain deficiencies in Castells's work. To illustrate the cen-tral theme of this stage – that is, of the intrinsic structural connections between gender and production relations – empirical material on the changing nature of women's work in Britain is presented. I conclude by positing a probable fifth stage of development, with some observa-tions for future research.

STAGES OF DEVELOPMENT OF
URBAN-SOCIAL THEORY

Much of Castells's urban-social theory, while not progressing in any linear fashion, can be recognized as having three main and distinct stages. This is not to imply that all of his research has followed these stages or that several stages are not present in any one work;[2] a brief overview of the content of Castells's research, followed by a discussion of its defining characteristics and of the construction of knowledge at each stage, will illustrate these points.

During the last decade the concept of urban-social change has been the subject of extensive debate, especially among those radical geographers and sociologists who have been strongly influenced by Castells's work. Attention has been focused particularly on the processes underlying the transformation of sociospatial forms and the creation of new urban meanings. Previously cities had often been classified as spatial units of production, according to the industries located within their boundaries or hinterlands. This approach presupposed a direct connection between local productive activities and the sustenance of the local, urban population. Although cities remain the places where the great majority of the population lives, they are no longer the circumscribed spatial unit within which productive activities are organized; the organization of production has expanded to regional, national, and international scales. In other words cities are embedded in, and sustained by, economic systems that extend far beyond their boundaries (Bondi and Peake 1988). Other indications of urban transformations that have received attention include the changing form of conflicts and protest in advanced industrial countries. There has been a shift in emphasis in urban areas, from class conflict between the owners of the means of production and labour, to pluriclass and community-based struggles over, for example, slum clearance, urban redevelopment, and the distribution of services including housing, health care, education, and transport. Such services, until the last decade and particularly within the United Kingdom and the United States, have been provided increasingly by the state.

Observations on the changing nature of political struggle and economic organization, and on the increasing intervention of the state into everyday life, led Castells to advance the persuasive argument that it is now more appropriate to consider the city as the locus, or spatial unit, of the "reproduction of labour power" rather than as a spatial unit of production. Castells initially equated the reproduction of labour power with collective consumption (of the state provision of goods and services), assuming that this formula was sufficient to cover the processes that enable people to return to work each day and to be replaced from one generation to the next. He argued that conflicts over collective consumption would effect structural change only if incorporated within the aims of a revolutionary political party; then, and only then, would the various organizations involved in these conflicts become urban-social movements.

As Castells has moved away from Althusserian-based Marxism and theoretical determinism to a concern with building up theory from empirical observation, his perspective on urban-social change has changed (Castells 1985a). In *The City and the Grassroots* (1983a) he argues that there are a number of sources of urban-social change, and the view that class conflict is the dominant source of that change is recognized as narrowly economistic. Class per se is not, however, dismissed: for Castells the dialectic of class restructuring and class struggle remains a major determinant of urban-social change, as are nonclass based social movements, including the feminist movement. But Castells's primary concern still lies with urban-social movements, which, he claims, now work to redress three areas of potential social conflict: collective consumption, the goals of independent political action, and cultural identity. The latter two are realized through participation in local institutions and through territorial control, respectively. It is the historical specificity of urban-social movements in transforming the city and creating new urban meanings that remains central to Castells's thesis; underlying much recent research is the notion that the variety of processes involved in the reproduction of labour power is the key to understanding the structure of, and activity within, urban areas. Certain features of this overview of Castells's work will now be treated in more detail and specified by stage in order to examine the development of urban-social theory.

Male Urban-Social Theory

In this stage of research knowledge is commonly shaped by an androcentric (i.e., male-as-norm) view of the world. Women's absence from theories of urban-social change is not recognized, nor is it stated that women's and men's experiences are the same; women are simply not considered. At worst this purportedly gender-neutral knowledge – supposedly based on science, truth, and objectivity – has been instrumental in perpetuating women's subordination, because the compartmentalization of knowledge into traditional, male-oriented subjects has prevented women from asking questions about their own existence (Spender 1982).

This stage is best illustrated by Castells's early work on urban-social change, exemplified by "Is there an urban sociology?" (1976a; see also

Castells 1976b, c). The article commences with a brief survey of urban studies, which Castells divides into three areas: urbanization as a worldwide process; studies of social disorganization and acculturation; and the "community study" tradition. The extent to which the city is defined in relation to men is suggested by the following extract: "the city is the product of History, the reflection of society, the action of Man upon space as he constructs his abode." Thus one is reassured that urban sociology is linked to the future of mankind. And indeed, how can one take exception to such wise remarks? Castell goes further: "One is obliged to accept the good sense of such a general statement" (Castells 1976a, 41–2). Given the embryonic status of feminist studies in the late 1960s, perhaps Castells cannot be criticized for being merely a product of his time. Yet six years later, he appears to remain unaware of the extent to which male experience is only a particular segment of all possible experience. In "Towards a political urban sociology" (1977a) he states: "While we mean by urban, a certain style of society ... we also mean by this term a certain social organization of space characterised by the concentration and interpenetration of man and his activities" (62). That Castells is not using "man" in a generic sense (encompassing both sexes) is made clear when he goes on to criticize liberal urban analyses for their concentration upon those actors, "each one of whom is defined by his attempt to maximise his power and gains. Thus, trade unions, the press or businessmen are considered to be on the same level" (64). Castells's concern is obviously with male members of the public worlds of formal politics, waged work, and institutionalized power. Furthermore not only are the issues addressed those of men, but knowledge for Castells is constructed from a male perspective and, moreover, is valued as the only knowledge worth having.

In the preface to *The Urban Question* (1977b) Castells makes clear his thoughts on the construction of knowledge: "What is not acceptable ... is that social practice should directly determine one's treatment of questions. Between the questions and the answers there must be a relatively autonomous mediation: the work of scientific research" (viii). Given that scientific research at this time for Castells was synonymous with the reconstruction of urban sociology along Althusserian lines, it is not surprising that he ignores questions of women and gender and examines issues solely in terms of class analysis. With the belief that there is an epistemologically superior mode of production of knowledge – Marxism à la Althusser – reality is seen as being ultimately constituted by one, privileged set of social relations (namely, those of class or accumulation). Any other social relations that can form the organizing principles of society are subsumed within the overarching theoretical category of production, empirically interpreted as male waged work under which all other experience, values, and understandings of the world are subordinated.

Compensatory Urban-Social Theory

In this stage women's absence is recognized and they are admitted into the analysis. References to women, however, tend to be brief, and consist of a few sentences or a paragraph; frequently they are labelled with one of several general descriptive variables. Hence little is known about women, and what is known is categorized according to criteria stipulated by men; an androcentric view still dictates the frame of reference. The patriarchal framework for the construction of knowledge is upheld by the male monopoly on meaning. Given male definitions of the world and of work, women are denied the reality of their own lives. For example women who often spend more time on housework than their male partners spend doing waged work are still thought of solely as housewives – as people who do not work. Hence, although women's activities and needs might be addressed within this stage or approach, they are presented – because categories of production have not been reformulated – as some form of special need, child care, for example, is considered to be women's problem and not an integral component of the relationship between the family-household system and the capitalist mode of production. So, although women gradually have been added to the record, the patriarchal content of and the construction of knowledge in urban-social research are not challenged at this stage.

This stage is well illustrated by case studies common to two of Castells's works, *The Urban Question* (1977b) and *City, Class and Power* (1978). In the former, a Paris-based study of the struggle for rehousing (in an area Castells refers to as Cité-du-Peuple), tables outline the socio-economic characteristics of residents and their conditions of residence. One of the variables relates to the percentage of active women in each study area, but there is little mention of them in the text; reference is made instead to genderless tenants, residents, and individuals. The one mention of women, in the Rue de la Boue example, counterposes them to militants: "One day, workmen arrived to cut off the water. There was a general mobilization. The militants were there. But all the housewives in the street were there too" (1977b, 342). Women may be "there," but not in their own right: women appear as the "other," as deviants to the male norm of radicalness. The women are viewed only in terms of their traditional reproductive roles, and their presence is interpreted as being supportive of the male militants and as an extension of their domestic role in the family. Later in *The Urban Question*, in a table summarizing the case study, the social force in the Rue de la Boue is labelled as consisting of "tradesmen" and some "proletarianized students"; the housewives thus disappear altogether.[3]

For Castells the effect on the lives of the women involved in these activities is not an issue for consideration, because ultimately protests are viewed only as male activities. Not taken into account, for instance, is the fact that for the housewife there is no separation in space, time, or identity between work and rest; or that the home, the street, and the community as workplaces can become the locus of political activity for issues of reproduction in a manner that parallels the function of the shopfloor in connection with production issues (Mackenzie and Rose 1983). Seen only as housewives, women are relegated to the domestic sphere beyond the bounds of the political; women, it seems, may be perceived as being part of the social structure, but not as being part of the structure of power (Rendel 1981).

Bifocal Urban-Social Theory

In the bifocal stage of urban-social theory, the gradual recognition of various facets of women's lives is also accompanied by an attempt to put forward positive images of women. Women's activities, however, tend to be studied in a dualistic manner, and the focus is on either their engagement in waged work or their work in the home, but rarely on both. Moreover it remains that women's lives still are being viewed through male eyes: it is women's waged work that needs to be "explained," while men's waged work is taken for granted. And no questions are asked about men's activities in the domestic sphere, which is unequivocally seen as women's sphere and disconnected from the world of men's and women's public employment. While there are also preliminary attempts to recognize the roots of women's oppression, and the attempts of women to tackle it, there is still an emphasis on a dualistic conceptualization of knowledge, experience, and action: male versus female; production versus consumption; objective versus subjective; rational versus emotional; public versus private; structure versus agency. Male and female spheres of life and individual qualities are perceived as complementary but separate. Overall women are included in this stage of urban-social theory if they have entered the public world of men, where they are seen to be collectively engaged in masculine-defined activities.

In Castells's work this stage is best illustrated by *City, Class and Power* (1978), in which he has come to recognize the power of women and their role in socialized consumption processes:

For example the feminist movement is threatening the very logic of the urban structure, for it is the subordinate role of women which enables the minimal "maintenance" of its housing, transport and public facilities. In the end if the system still "works" it is because women guarantee unpaid transportation (movement of people and merchandise) because they repair their homes, because they make meals when there are no canteens, because they spend more time shopping around, because they look after others' children when there are no nurseries, and because they offer "free entertainment" to the producers where there is a social vacuum and an absence of cultural creativity. If these women who "do nothing" ever stopped to do "only that" the whole urban structure as we know it would become completely incapable of maintaining its functions. The contemporary city also rests upon the subordination of "women consumers" to the "men producers." (1978, 177)

The dualistic categorization of women as consumers and men as producers is the basic characteristic of this stage. Not only is women's role as consumer seen as natural, it is also seen as their only role. Furthermore because women are not seen as producers (waged workers), they cannot be exploited but only subordinated. Hence when Castells refers to the social relations of exploitation and domination that result from the material organization of daily life by capital and by the state, women are, by default, excluded from the analysis. Castells's conceptualization of women solely as consumers and of men solely as producers is remarkably inaccurate: gender roles cannot be mapped directly on to a dichotomy between production and consumption (Bondi and Peake 1988).

Perspectives associated with this stage can also be glimpsed in Castells's later work, *The City and the Grassroots* (1983a), in which it appears he has gone beyond treating women and sex as simple, natural facts: "How do class, sex, race ... contribute to the formation of social actors that intervene in the urban scene?" (1983a, xviii), he asks in the preface; and in the book there are a number of historical case studies on urban-social movements that investigate the role and nature of women's participation. In three detailed studies – the 1915 Glasgow rent strike, a working-class struggle for the reproduction of labour power; the Paris *Commune* of 1871, a municipal revolution designed to improve the level of well-being of the people; and the 1922 rent strike in Veracruz, Mexico, a revolutionary populist movement – Castells recognizes the dominant and leading role played by women. Both the Glasgow and Veracruz rent strikes were organized and led by women. Indeed in Veracruz the rent strike was "a movement of women in a revolt rooted in their role as agents of organisation of all spheres of everyday life" (Castells 1983a, 46). And women's role in the Paris *Commune* was "crucial," because it included leading the mobilizations and joining in the demonstrations and fighting. Castells, however, claims that "[i]n none of the cases observed were the gender relationships at the core of the conflict, in spite of the decisive participation of women in the movements: a striking verification of our hypothesis about the historical hierarchy between relationships of production, power and gender" (Castells 1983a, 335). In this case determination of the form of urban-social change for Castells lies not in the intersection of the modes of production and reproduction but solely in the mode of production.

Castells does go on to recognize that the form of this hierarchical relationship is being called into question by the present-day women's movement: "the redefinition of urban meaning to emphasize use value and the quality of experience over exchange value and centralisation of management is historically connected to the feminist theme of identity and communication" (Castells 1983a, 309). But how can we know the historical role gender relations played, when the overall emphasis in his case studies is on sex, not gender – on male versus female, with women's activities discussed separately and only *after* the major determinants of the movements have been decided? The social profile of the women in the Paris *Commune* case relates solely to the family situation; we are told only the occupations of the men. Although in the Glasgow rent strike study we are told that the women were industrial workers, their occupations are not expanded upon. In addition we are told that some were housewives, widows, and suffragettes. In the Veracruz rent strike reference is made only to women as prostitutes and "common women." Fundamentally women still are seen to occupy a separate sphere; women still are the consumers, and men, the producers: "Throughout history male domination has resulted in a concentration and hierarchy of social tasks: *production*, war, and political and religious power – *the backbone of social organization* – have been *reserved for men*. All the rest, that is, the immense variety of human experience, from the bringing-up of children and domestic work to sensual pleasure and human communication, have been the women's domain" (1983a, 68; emphasis added).

Despite recognizing the vast area of activities covered by women, Castells still sees their most salient role as that of consumer. In referring to the Glasgow women rent strikers he asserts, "They claimed the right to live for their families and they were the agents of a consumption oriented protest, as a continuation of their role as consumption agent within the family, even when they were workers at the same time" (1983a, 31). The tendency still is for Castells to privilege production as a male preserve, characterized by wage employment outside the home. This theoretical deficiency appears to arise from the fact that Castells has maintained his bifocal view of women and, therefore, has insufficiently problematized gender relations, which he treats solely as a category that helps to make sense of the particular ways in which people transform and experience urban life. While he also claims to see gender as a relational category of inequality, as a set of spatially and temporally variable social processes that enter into all other social relations and therefore partially constitutes them, in effect he views social relations as hierarchical rather than interdependent.

In subsequent work, which has focused on the role of technology in structuring socio-spatial forms, Castells (1983b, 1985b, d) has started to deconstruct his androcentric concept of social change. In place of his economistic explanations of urban-social change is a dialectical approach centred around collective consumption, political self-management, and cultural identity; he recognizes that it is gendered human agents who create, maintain, and transform structures in their struggles to create new urban meanings. However Castells continues to ignore an important dimension of these struggles: the changing forms and relationships of women's work in advanced capitalistic societies, along with the social and spatial implications of these changes. The absence of this dimension is evident by the treatment of the home as a vacuum that is counterposed to the workplace. The impact of high technology, for instance, according to Castells, is likely to lead to "a decrease in functional travelling around urban areas and a concentration of activities around three major poles: work places, homes and pure leisure places" (1985b, 17). While Castells recognizes that urban form will be shaped by processes other than technology, such as ethnic cultures, local networks, and community organizations, he makes no reference to potentially divergent trends concerning the adoption of high technology. The take-up of high technology in the home – what Castells refers to as the "home information revolution" – is not merely a function of high technology itself but rather will depend on the logic of certain developments that affect power relations in the family-household system (including increases in male and female unemployment, the upward trend in homeworking, and so on). Castells (1983b) recognizes the growth of homework as a major tendency in the newly emerging spatial division of labour: "the process of work is shrinking in space and is becoming domestic again" (1983b, 18), he writes, and indeed he views this process as "shaking the foundations of our social fabric" and leading to fundamental alternations in both "the process of work and the organisation of everyday life" (1983b, 18); yet he still treats the domestic arena as an unproblematic area of apparent autonomy, as opposed to a gender-differentiated domain.

Castells, therefore, does not address the highly specific gendered meaning of work in the home and fails to take into account either household social divisions that lead to the burden of women's "double shift" or the complex nature of the relation between women's paid and unpaid labour and the wage system.

Although the bifocal stage of research has served a useful purpose, its lacunae are becoming too great to ignore. With emphasis on the differences as opposed to the similarities between spheres, their analytical separation has appeared as natural. For example by insisting that production and reproduction occur in two different places, researchers ensure that different social relations characterize these processes to the extent that they continue to merit separate investigations. Moreover one set of social relations within each of these places – the waged workplace and the home – has been emphasized at the expense of all others, namely class and gender, respectively. Consequently within the home unpaid domestic labour is seen as the only type of work that takes place, whereas it is only one strand of the many relationships between the family-household system and production. The reality of the everyday lives of women and men is such that they are involved in both reproduction and production (which may or may not be part of the wage relation central to the capitalist mode of production). And the reproduction of labour power is not a set of processes that take place solely in the home, just as production processes do not lie totally outside it. Thus arenas of production and reproduction are no more gender specific than are production and consumption. Rather gender differences are the product of the different ways in which production and reproduction processes combine within, and affect, the lives of women and men of all social groups (Bondi and Peake 1988).

To summarize briefly, Castells's work has stopped short of a feminist stage of development by failing to recognize the various ways in which the family-household system is linked with production. Since Castells does not view women in their entirety, as both producers and reproducers, he does not attach sufficient importance to the dependence of capitalism on women's unpaid domestic labour for reproducing the labour force, nor to the fact that the conditions for the reproduction of the family-household system are shown to lie partly outside itself – all of which is vital for an accurate assessment of the status of women's economic contributions and for a fuller understanding of the interaction between class and gender in the workplace, and its impact on urban-social change.

THE FOURTH STAGE: FEMINIST
URBAN-SOCIAL THEORY

As a number of feminist critiques have asserted (see, for example, Bowles and Duelli-Klein 1983; Crompton and Mann 1986; Delphy 1984; Kuhn and Wolpe 1978; Mies 1986; Spender 1982; see also the chapters by Bondi, Kobayashi, and Christopherson in this volume), the categories of bifocal analyses are in many cases inappropriate and fail to fit the reality of people's everyday lives, particularly women's. Bifocal urban-social theory has enabled us to gain useful insights into the structure of society, in terms of both spatial criteria (such as the physical divide between home and workplace that has grown up since the industrial revolution) and social criteria (for the public world of men, politics, and production and the private world of women, community, and reproduction). But feminist theory has shown that this dualized conceptualization is no longer valid; evidence for the separate spheres theory is far from convincing and, when subjected to empirical investigation, these analytical categories have been shown to be not mutually exclusive, fixed spheres but shifting, dynamic domains, the boundaries of which are both elusive and culturally and historically specific.[4] Hence feminist academics' concern with exploring the nature of women's exploitation has led not only to the question of what is specific and distinctive about women's relation to the processes of production and reproduction, but also to the question of whether these concepts are the most cogent for expressing the social organization of everyday life (Peake 1986b). Castells's preoccupation with consumption, particularly collective (or socialized) consumption, for example, has left a gap in research on the interaction of processes of production and reproduction; and his reduction of the reproduction of labour power to collective consumption has excluded aspects of daily life that are of particular importance to women (Wilson 1977).

For women the domestic sphere has significance as a manifestation of their central role in the reproduction of labour power. Women, in conjunction with the state (particularly at the local tier), provide the services that daily and generationally reproduce labour forces and the social conditions of production. Women are also centrally involved in other aspects of the reproduction of labour power, including biological reproduction, the socialization of children, cooking, cleaning, and so on. The belief that women are by nature destined to provide such services is reinforced daily by their concrete experience: servicing is a material condition of their lives. The reproduction of labour power, apart from state provision, also involves private sector provision, domestic servicing activities, and biological and human reproduction. But these aspects of reproduction tend to be viewed, bifocally, in a biologistic manner: they are labelled as "natural" without being appreciated as the production of the means of existence. A broad conception of women's activities, then, would include those that maintain human life: those that either must be done if people are to survive, or those that, if done, lead to improved living conditions or a greater sense of well being. This definition is concerned not only with reproductive or servicing work (variously referred to as community managing, community care, or informal caring networks) but also with waged work. These concerns are now taken up in an examination of the nature of women's work in Britain.

There have been a number of interlocking postwar social and economic transformations in Britain that have changed the nature of work, and in which the articulation between gender and production has been quite clearly revealed (Mackintosh 1981).[5] These include the processes of industrial restructuring; the concomitant growth of the service sector economy and the increase in the number of women in waged work; the increase (until the last decade) in the role of the state in everyday life, particularly in the provision of services and the growth in public sector occupations; the rise in unemployment (up to the mid-1980s) and the increase in part-time employment; and the evolution of family structure away from the traditional norm of the nuclear family to a number of diverse forms. Further with the global recession of the late 1970s and early 1980s there have been a number of additional economically driven changes (see, for example, Allen and Wolkowitz 19877; Purcell et al. 1986; Redclift and Mingione 1985), including a decline in the authority of previously powerful institutions, such as trades unions; the emergence of political groups with increasingly divergent ideas, such as environmentalist groups and the antinuclear lobby; the removal of regulatory mechanisms by government and an increase in privatization; and a casualization of the labour force, with an increasing diversity of working patterns and relationships.

From the postwar period in Britain up to the late 1970s, there was a sustained growth of women's participation in the labour force.[6] Between 1951 and 1971, 2.5 million jobs were created, of which 2.2 million were taken by women, including, at least initially, many arriving from the West Indies and East Africa. Whereas in 1951 women accounted for 30 per cent of the labour force, by 1981 they accounted for 42 per cent and reached a peak of 10.2 million (Women and Geography Study Group 1984). The overriding feature of this growth in women's participation in the labour force was the increase in the proportion of married women, who worked largely in part-time jobs in the service sector. Postwar expansion in women's employment occurred principally in the state sector: the new jobs were for clerical workers, nurses, ancillary health workers, teachers, social workers, and so on. The growth of female participation in waged employment itself fuelled the demand for services such as nurseries, school meals, and homes for the elderly, while the wages earned by women stimulated the production of consumer goods and services, including convenience foods, automatic washing machines, and launderettes.

The entry of women into the work force has also been associated with changes in family life. Notwithstanding demographic variations between racial and social groups, both completed family size and the time between the last birth and a woman's return to waged employment have, on average, decreased in recent years. A woman's life in waged employment is thus becoming longer and, in this respect (aside from part-time work), looking more like a man's. The age of the youngest child is the major determinant of whether women work outside the home and whether they work part- or full-time (Beechey 1986). It is, therefore, the presence of children rather than marital status that determines women's labour market participation. Figures for 1980 reveal that 65 per cent of all women in Britain were economically active. This figure breaks down to 72 per cent for all nonmarried women and 62 per cent for married women, with this latter figure rising to 71 per cent by the time the youngest child is ten years old (Martin and Roberts 1984).

These developments have not necessarily eradicated preexisting gender divisions, but they have resulted in the need for women to divide their time among more activities. Wilson (1977) has argued that the growth of the welfare state in Britain did not relieve women of their primary responsibility for domestic life; in fact it increased the role of the state in perpetuating existing gender divisions. For example after World War II, state provision of child care for young children was almost entirely withdrawn. Despite substantial demand, state provision for pre-school-aged children has remained limited, and a 1988 European Commission report revealed that British child care provision was the worst in Europe (Judd 1988). Thus the entry of women into employment has altered but not reduced the role of women as reproducers: women continue to be on the "front line," fighting for the provision of services by the public sectors, frequently in a vicarious fashion on behalf of other members of their households. Many women can aptly be described as having dual roles, as wage earners and as unpaid domestic service workers; women's performance of domestic work not only expresses their subordination in the family (since men actively benefit from this work), it also weakens their position in the waged labour market (Mackintosh 1981) – as is borne witness to by the prevalence of part-time working women.

The participation of women in the labour force has not paralleled that of men in other respects. Despite the 1970 Equal Pay Act and the 1975 Sex Discrimination Act, women's earnings remain markedly lower than those of men: the average hourly pay for women is approximately 75 per cent that for men. The failure of the sex equality legislation to achieve more success is largely a function of gender segregation in occupations. Of men in the labour force, half are in jobs where the work force at their place of employment is 90 per cent male; of women, half are in jobs where the work force is 70 per cent female (Phillips 1983). Women's jobs typically involve tasks requiring supposedly female attributes such as dexterity (for component assembly in the electronics industry), docility (for routine clerical and cleaning work), and caring skills (for nursery teaching and nursing). Most of these tasks are classified as unskilled or semiskilled and most relate closely to those required for domestic servicing work (Hakim 1981; Beechey 1986). Thus the position of women in the work force can serve to reinforce the ideology of domesticity rather than to undermine traditional divisions of labour within the family.

Despite the significance of gender divisions at places of waged work and within the family, it must not be assumed that men and women form homogeneous groups. Gender divisions exist alongside, and interact with, class, racial, and ethnic divisions (Anthias and Yuval-Davies 1983). For example rates of economic activity vary between women of different ethnic groups: in 1981, 23 per cent of White women, 47 per cent of West Indian women, and 25 per cent of Asian women were in full-time work; 17 per cent of White women, 14 per cent of West Indian women, and 5 per cent of Asian women were in part-time work (1981 Labour Force Survey). The high economic activity rate of West Indian women can be attributed partly to the fact that in the 25–34 year age group West Indian women's participation in the labour market increases, whereas that of White women decreases (Beechey 1986). The figures for part-time work also verify that it is White (married) women who are more likely to work on this basis, while women from ethnic minorities are more likely to work full-time. Moreover gender segregation at work is inseparable from racial segregation. White women are concentrated in nonmanual occupations (especially clerical work and retailing); Black women, in manual work; Asians, particularly in the textiles and clothing trade (especially in Yorkshire and the East Midlands); West Indian women, in engineering and the health services (see Mama 1984). The majority of women, regardless of race, are in low-level jobs, but this is particularly true of West Indian women, of whom only 2 per cent are in professional and managerial jobs, compared with 17 per cent of White women and 20 per cent of Asian women (Beechey 1986).

The current recession in Britain, which began in the late 1970s, has further changed the waged work and domestic activities of women. Cuts in state provision affect everyone but, as Edgell and Duke (1983) have noted, "the current attack on the welfare state is predominantly an attack on women as a) the major contributors of waged labour to collective social provision, b) the major beneficiaries of collective social provision, and c) the major care role providers in the family and society" (375). Women are the major users of services, both for themselves and for others, particularly children. In 1982 the Equal Opportunities Commission estimated that there were approximately one and a quarter million carers in Britain, most of them women (EOC 1983). In London, for example, it has been estimated that not only are women in full-time employment sixty-six times more likely than a man to have to care for a sick child, but there are also more women caring for elderly and disabled relatives than for children under the age of sixteen (Women's Equality Group 1987). Consequently some women have to give up work to look after the young, the old, or the infirm, while others have lost jobs as services have been cut. Overall, job losses have been greatest in regions dominated by traditional industries such as shipbuilding and steel, which have always been male preserves. In these areas women's employment has remained relatively buoyant (Massey 1984). Women suffer disproportionately under "last in, first out" redundancy policies, however, and from practices of dismissing part-time staff, who have least employment protection. Thus between 1976 and 1983, unemployment among women is estimated to have increased fourfold (Beechey 1986).

Again these trends have affected different groups of women to differing degrees. In terms of unemployment, rates are higher for Black men and women of all age groups than for White men and women. The incidence of low pay is also greater for non-Whites than for Whites. For example of the 750,000 women who work for the National Health Service, the majority are in low-level jobs: 75 per cent of the ancillary workers are female and many are Black women with wages below the official poverty line (Doyle 1987). Nationally it has been estimated that two-thirds of the jobs lost as a result of privatization in the National Health Service have been women's jobs, and in London between 50 to 80 per cent of these had been occupied by women from ethnic minorities (Women's Equality Group 1987; also see Bryan et al. 1985). Moreover divisions between men and women, between racial groups, and between social classes have tended to make the implementation of policies that result in job losses easier (see Carby 1982; Mama 1984; A. Phillips 1987).

High rates of unemployment have been accompanied by a polarization within the work force, separating those in secure and better-paid jobs, for whom living standards have risen, from those in insecure and low-paid jobs. The latter group includes approximately one in five male workers and two out of three female workers (Bazen 1985). Consequently for a substantial sector of the population, financial resources have declined as the demands on the family have increased. Women in employment have not only been hit by the recession in terms of job loss by way of early retirement, redundancy, or (in the case of school leavers) simply failing to find a job; but those who remain in employment are finding their jobs restructured and redefined, through new technology agreements, for example, which results in new divisions of labour (both within and between jobs). The most evident change in women's employment in Britain over the last decade has been the increase in the number of part-time workers. Between 1975 and 1985 male employment fell by 12.9 per cent, full-time female employment by 8.4 per cent; but the number of women working part-time increased by 22.3 per cent (E. Phillips 1987). By 1985, 47 per cent of women in waged work (and 55 per cent of married women) were part-time workers, and 88 per cent of all part-time employees were women (Beechey 1986). Over 80 per cent of women in part-time work are employed in the service sector – in distribution, in financial, professional, and scientific services, and in public administration – and this figure is increasing. There has also been an increase in the casualization of employment, leading to modifications in workers' statutory rights and benefits. Women who work less than 16 hours a week, for example, are deprived of their statutory rights to maternity leave, overtime and holiday pay, pensions, and unemployment and sickness benefits. It is also harder for part-timers to qualify for benefits: many have to work three years longer for the same employers than do full-timers in order to qualify for basic rights. And there have been increases not only in part-time work but also in short-term and fixed temporary contracts (not only for manual workers, but also, for example, for office workers, teachers, and nurses), in self-employment, and in homeworking.

Estimates of the number of homeworkers are difficult to assemble but there appears to be agreement with Hakim's (1984) figures that one in seven women in waged employment works from her home. Given that in 1983 there were 8.8 million women employees (EOC 1983), then there are approximately one and a quarter million female homeworkers in Britain. Convincing arguments have been made to indicate that this number is increasing (see, for example, Allen and Wolkowitz 1987 and Huws 1984). They are often engaged in such tasks as making clothes, finishing off shoes, filling Christmas crackers, sewing buttons on cards, and checking items for flaws. Since many are working in industries not covered by minimum hourly rates, they are paid below the legal minimum; a recent survey by the National Home-working Unit found that 90 per cent of homeworkers were underpaid. The government, however (as of June 1988), had no plans to create separate legislation for homeworkers, claiming that they already have the same protection as other workers – despite the fact that the Health and Safety Acts, for example, do not cover private homes (Harris 1988).

Homework is defined by Allen and Wolkowitz (1987, 1) as "the supply of work performed in domestic premises, usually for piecework payment." They recognize this as only one of seven types of work carried out in the home (26):[7] unpaid domestic labour; work for household provisioning (e.g., growing and preserving fruit and vegetables); do-it-yourself (DIY) home maintenance and improvement; reciprocal exchanges of goods and services between kin and neighbours (e.g., baby-sitting, repairs, exchange of tools, etc.); illegal work (e.g., work carried out by children or criminal activities); informal work unrecorded by the employer or self-employed (e.g., unregistered child-minding, window-cleaning, commission selling, etc.); homework (or outwork). As a form of commodity production homework differs from other work carried out in the home in that it enters into the wage relation that is central to capitalist production. As part of the capitalist labour process homeworking is only one of several instruments of industrial restructuring (such as flexible "manning," fragmentation of the labour process, privatization, occupational segregation, self-employment), whereby the relocation of production (in this case from the factory or office to domestic premises) can result in the casualiza-tion of employment. The above list reveals the extensive number of forms of work outside of this wage relation that exist alongside capitalist production. An understanding of homeworking as an increas-ingly common form of capitalist production that takes place in an arena not characterized by commodity production, but that forms a link between capitalist and noncapitalist production, may help us to reformulate the framework within which work is analyzed. Women's experience of work does not always conform to the male pattern: they do not always engage in waged work outside the home for a set number of hours each day and week, or over a lifetime, and they have a wide variety of workplaces where they exercise their skills in caring for, or servicing, other people. Thus when we consider the type of work done by women, we see it covers activities of both production and reproduction, both inside and outside the home. The type of work that women do, as well as the social relations of production under which this work is performed, necessitates that we rethink these categories.

FUTURE RESEARCH

A feminist analysis of urban-social theory has enabled us to reach a stage that has required us both to cross over disciplinary boundaries and to recognize our limitations as products of a patriarchal society. Research at this stage is concerned with treating women as subjects, not objects, and with women themselves asking the questions. Not only have new questions been asked, but new categories and concepts have been developed through which we can view women's needs, values, and experiences (Peake 1986a). Investigations of the reasons that women themselves have given for moving in and out of waged and domestic work, in place of those concerned solely with macro-economic factors of supply and demand, have led to the recognition that women's consciousness has been evolving – not only women's individual consciousness of their roles, of what it means to be a woman, but also a collective feminist consciousness. The development of this feminist consciousness has allowed our understanding of gender relations to be questioned. How are they constituted and experienced? How do we think about them? How can we reevaluate and alter them? As a result, not only have patriarchal frameworks been attacked, but the very nature of the social sciences has been questioned. Male control has been challenged, as the dominated as well as the dominators have become legitimate subjects for study.

Are we now moving into a stage of multifocal, relational research, through which we can develop an holistic knowledge that accounts for both female and male experience and that can be used for the empowerment of people to transform the real world? With the recognition that knowledge (and self) are socially constituted – that our mode of understanding depends upon our social practices and cultural contexts – different views of the social construction of reality, of how we live and behave and make sense of what we can do, make us question whether we all share the same social meanings and experiences. Studies of women form only part of this stage's attempt to understand and reformulate gender relations. There is also a concern to reveal common aspects of female and male experience and in addition to focus on particularity, on the nodal points where women's and men's lives intersect (Thompson-Tetrault 1985). Academics are only beginning to formulate the necessary concepts and to collect the data that will enable us to look at the processes of continuity and change in people's everyday lives (see, for example, Dwyer and Bruce 1988; Mackenzie 1988a, b; Pratt and Hanson 1988; Wolch and Dear 1989). In particular, comparisons of the lives of women and men have hitherto been set in the public sphere concentrating on their different employment experiences. We know little about the ways in which they interact in the domestic sphere. The social relations of the domestic domain and their historical and cultural variations form a major gap in our knowledge, the bridging of which would help to establish a continuum between women's and men's experiences. Any account that falls short of such a comprehensive approach runs the risk of taking only a partial view of both urban-social change and women's and men's lives, of ignoring men and of viewing women as objects or, even worse, as victims, and of failing to appreciate both the interactions between men's and women's roles and experiences and the rationality and creativity expressed in their strategies for coping.

CONCLUSION

The central aim of this essay has been to illustrate what a feminist perspective can add to our understanding of urban-social change. Castells's analyses of urban-social change attempt to define what is historically specific about those structurally conditioned social processes through which people struggle to impose their collective meanings upon city and society. These very struggles to create new urban meanings not only produce urban change but also result in the redefinition of the participants. One of the major arenas where new meanings are being created is the world of work, but in Castells's analyses there is a tendency to ignore the topic of women and work. Having transformed his perspective from one that focused on a gender-blind concept of production and an economistic concept of reproduction to a concern with the gender-aware politics of reproduction, Castells has devised a theory of urban-social change that does not address women in their entirety as both producers and reproducers. While spatial and social implications of the changing structure of work per se are addressed by Castells and others, important dimensions have been left unexplored, including the meaning of waged work to women's lives as well as the impact upon relations between women and men of the increasing role of the home as both a waged and an unwaged workplace. An overview of the changing nature of women's work in Britain illustrated that knowledge of this topic is essential for an in-depth understanding of the processes underlying the creation of new urban meanings.

The incorporation of women into waged work, both inside and outside the home, has significantly transformed existing gender relations. Of course gender relations will continue to evolve, whether equitably or inequitably, and these, in turn, will have a reciprocal effect on the transformation of urban environments. But the ways in which these environments are reconstituted, although temporally and spatially specific, will be determined not only by the outcome of struggle between women workers, male workers, capital, and the state but also by the relationship between women and men in the domestic sphere. I have argued that it is only by considering the totality of people's lives that we will develop a full understanding of the various practices integral to urban-social change and the reciprocal ways in which these changes are restructuring social relations. To this end we need to adopt a multifocal, relational perspective so that we can begin to compare women and men in all aspects of their lives and show that work-based activities are redrawing the boundaries of urban-social theory by bringing about a reunion of places of work and residence, thereby revealing gender to be a major determinant of urban-social change.

NOTES

I would like to thank Richard Harris, Linda McDowell, and Janice Monk for their comments on an earlier draft.

1 In choosing Castells's work my intent is not to criticize Castells himself; all work is dependent upon the social context within which it is produced, as Castells himself is at pains to point out. He states his preference (1985a, c) that his work be viewed not as a final product but as a tool to enhance understanding, which is the perspective I take here.

2 The phases of urban-social theory development presented here are drawn from Thompson-Tetrault's (1985) work on social theory in general (see also Bowles and Duelli-Klein 1983 and Spender 1982 for similar accounts). Whereas she discusses only the content and construction of knowledge within each stage, I also consider the purpose for which this knowledge is created.

3 In *City, Class and Power* the social force referred to includes the community and proletarianized students living in the area (1978, 120).

4 For a useful critique of the notion of separate spheres in relation to the Third World, see Moser 1981; Bromley and Gerry 1970; Moser and Peake 1987. For feminist critiques, see Garmanikow et al. 1983; Coward 1983; Molyneux 1979. And for feminist accounts of the validity of the spatial divide, see Bondi and Peake 1988; Mackenzie 1988a; L. McDowell

1983; Matrix 1984; Wekerle 1984; and the Women and Geography Study Group 1984.

5 Within this national overview women's position in the labour market obviously varies from place to place, and reflects the extent to which women's employment opportunities are dependent upon local labour-market conditions. An analysis of the increase in women's employment in the postwar period reveals distinct regional variations, matched by similarly distinct spatial patterns of job loss (see Walby 1986).

6 See Bondi and Peake (1988), from which the greater part of this account of the changing nature of women's relation to production and reproduction is taken.

7 Pahl and Wallace (1985) pose an alternative classification constituting three categories: wages and salaries exchanged for labour; self-provisioning (i.e., the production and consumption of goods and services by household members for other members); and activities carried out by members of other households for exchange. Thus they include homework in the first category and treat it in the same manner as employment in, for example, an office or factory.

REFERENCES

Allen, S. and C. Wolkowitz. 1987. *Homeworking*. London: Macmillan.

Anthias, F. and N. Yuval-Davies. 1983. "Contextualising feminism-gender, ethnic and class divisions." *Feminist Review* 15: 62–75.

Balbo, L. 1987. "Crazy quilts: rethinking the welfare state debate from a woman's point of view." In *Women and the State*, A. Showstack-Sassoon, ed, 45–71. London: Hutchinson.

Bazen, J. 1985. *Low Wages, Family Circumstances and Minimum Wage Legislation*. London: Policy Studies Institute.

Beechey, V. 1986. "Women's employment in contemporary Britain." In *Women in Britain Today*, V. Beechey and E. Whitelegg, eds, 71–131. Milton Keynes: Open University Press.

Bondi, L. and L. Peake. 1988. "Gender and the city: urban politics revisited." In *Women in Cities: Gender and the Environment*, J. Little, L. Peake, and P. Richardson, eds, 21–40. London: Macmillan.

Bowles, G. and R. Duelli-Klein, eds. 1983. *Theories of Women's Studies*. London: Routledge and Kegan Paul.

Bromley, R. and C. Gerry, eds. 1970. *Casual Work and Poverty in Third World Cities*. New York: Wiley and Sons.

Bryan, B., S. Dadzie, and J. Scarfe. 1985. *The Heart of the Race*. London: Vigaro.

Carby, H. 1982. "White woman listen! Black feminism and the boundaries of sisterhood." In *The Empire Strikes Back: Race and Racism in 1970s Britain*,

Centre for Contemporary Cultural Studies eds, 212–35. London: Hutchinson.

Castells, M. 1976a. "Is there an urban sociology?" In *Urban Sociology: Critical Essays*, C. Pickvance, ed, 33–59. London: Methuen. Orig. pub., 1968. "Y a-t-il une sociologie urbaine?" *Sociologie du Travail* 1: 72–90.

– 1976b. "Theory and ideology in urban sociology." In *Urban Sociology: Critical Essays*, C. Pickvance, ed, 60–84. London: Methuen.

– 1976c. "Theoretical propositions for an experimental study of urban social movements." In *Urban Sociology: Critical Essays*, C. Pickvance, ed, 147–73. London: Methuen.

– 1977a. "Towards a political urban sociology." In *Captive Cities*, M. Harloe, ed, 61–78. London: Wiley & Sons.

– 1977b. *The Urban Question*. London: Edward Arnold. Orig. pub., 1972. In *La Question Urbaine*. Paris: Maspero.

– 1978. *City, Class and Power*. London: Macmillan.

– 1983a. *The City and the Grassroots*. London: Edward Arnold.

– 1983b. "Crisis, planning and the quality of life: managing the new historical relationships between space and society." *Society and Space* 1: 3–21.

– 1985a. From the urban question to the city and the grassroots. Urban and Regional Studies, working paper 47. University of Sussex.

– 1985b. "High technology, economic restructuring and the urban-regional process." In *High Technology, Space and Society: Urban Affairs Annual Reviews* 28, M. Castells, ed, 11–40. London: Sage.

– 1985c. "Commentary on G.C. Pickvance's 'The rise and fall of urban movements ....'" *Society and Space* 3: 55–61.

– ed. 1985d. *High Technology, Space and Society: Urban Affairs Annual Reviews* 28. London: Sage.

Coward, R. 1983. *Patriarchal Precedents*. London: Routledge and Kegan Paul.

Crompton, R. and M. Mann, eds. 1986. *Gender and Stratification*. London: Polity Press.

Delphy, C. 1984. *Close to Home: A Materialist Analysis of Women's Oppression*. London: Hutchinson.

Doyle, L. 1987. "Carers and the careless: the prospect for the Health Service under the Tories." *Feminist Review* 27: 49–54.

Dwyer, D. and J. Bruce, eds. 1988. *A House Divided: Women and Income in the Third World*. Stanford: Stanford University Press.

Edgell, S. and V. Duke. 1983. "Gender and social policy: the impact of public expenditure cuts and reactions to them." *Journal of Social Policy* 12: 357–78.

EOC (Equal Opportunities Commission). 1983. *Seventh Annual Report: 1982*. Manchester.

Garmanikow, E., D. Morgan, J. Purvis, and D. Taylorson. 1983. *The Public and the Private: Social Patterns of Gender Relations*. London: Heinemann.

Hakim, C. 1981. "Job segregation: trends in the 1970s." *Employment Gazette* (November), p. 12.

– 1984. "Homework and outwork: national estimates from two surveys." *Employment Gazette* (January), p. 5.

Harris, A. 1988. "Hardship hides behind doors of homeworkers." *The Observer* (5 June), p. 7.

Huws, U. 1984. "New technology homeworkers." *Employment Gazette* (April), pp. 10–12.

Judd, J. 1988. "Britain's educational underclass." *The Observer* (10 April), p. 7.

Kuhn A. and A. Wolpe, eds. 1978. *Feminism and Materialism*. London: Routledge and Kegan Paul.

McDowell, L. 1983. "Towards an understanding of the gender division of urban space." *Society and Space* 1: 59–72.

Mackenzie, S. 1988a. "Building women, building cities: toward gender sensitive theory in the environmental disciplines." In *Life Spaces: Gender, Household, Employment*, C. Andrew and B.M. Milroy, eds, 13–30. Vancouver: University of British Columbia Press.

– 1988b. "Balancing our space and time: the impact of women's organisation on the British city, 1920–1980." In *Women in Cities: Gender and the Urban Environment*, J. Little, L. Peake, and M. Richardson, eds, 41–60. London: Macmillan.

Mackenzie, S. and D. Rose. 1983. "Industrial change, the domestic and home life." In *Redundant Spaces in Cities and Regions*, J. Anderson, S. Duncan, and R. Hudson, eds, 155–200. London: Academic Press.

Mackintosh, M. 1981. "Gender and economics: the sexual division of labour and the subordination of women." In *Of Marriage and the Market*, K. Young, C. Wolkowitz, and R. McCullagh, eds, 18–34. London: Conference of Socialist Economists.

Mama, A. 1984. "Black women, the economic crisis and the British State." *Feminist Review* 17: 21–35.

Martin, J. and C. Roberts. 1984. *Women and Employment: A Lifetime Perspective*. Department of Employment and Office of Population Censuses and Surveys. London: HMSO.

Massey, D. 1984. *Spatial Divisions of Labour*. New York: Methuen.

Matrix. 1984. *Making Space: Women and the Man Made Environment*. London: Pluto Press.

Mies, M. 1986. *Patriarchy and Accumulation on a World Scale*. London: Zed Books.

Molyneux, M. 1979. "Beyond the domestic labour debate." *New Left Review* 116: 3–27.

Moser, C. 1981. "Surviving in the suburbios: women and the informal sector." *Institute of Development Studies Bulletin* 12. 3: 54–62.

Pahl, R. and C. Wallace. 1985. "Household work strategies in economic recession." In *Beyond Employment*, N. Redclift and E. Mingione, eds, 189–227. London: Blackwell.

Peake, L. 1986a. "Teaching feminist geography: another perspective." *Journal of Geography in Higher Education* 10: 186–90.

– 1986b. "A conceptual inquiry into urban politics and gender." In *Politics, Geography and Social Stratification*, K. Hoggart and E. Kofman, eds, 62–85. London: Croom Helm.

Phillips, A. 1983. *Hidden Hands*. London: Pluto Press.

– 1987. *Divided Loyalties: Dilemmas of Sex and Class*. London: Virago.

Phillips, E. 1987. "Pushing back the tide: women in the public sector." *Feminist Review* 27: 43–7.

Pratt, G. and S. Hanson. 1988. "Gender, class and space." *Environment and Planning D: Society and Space* 6: 15–35.

Purcell, K., S. Wood, A. Wanton, and S. Allen, eds. 1986. *The Changing Experience of Employment: Restructuring and Recession*. London: Macmillan.

Redclift, N. and E. Mingione, eds. 1985. *Beyond Employment*. London: Blackwell.

Rendel, M., ed. 1981. *Women, Power and Political Systems*. London: Croom Helm.

Spender, D. 1982. *Invisible Women: the Schooling Scandal*. London: Writers and Readers Publishing Co-operative Society Ltd.

Thompson-Tetrault, M.K. 1985. "Phases of thinking about women in history: a report card on the textbooks." *Women's Studies Quarterly* 12.3, 4: 35–47.

Walby, S. 1986. *Patriarchy at Work*. London: Polity Press.

Wekerle, G.R. 1984. "A woman's place in the city." *Antipode* 16. 3: 11–19.

Wilson, E. 1977. *Women and the Welfare State*. London: Tavistock.

Wolch, J. and M. Dear, eds. 1989. *The Power of Geography: How Territory Shapes Social Life*. London: Allen and Unwin.

Women's Equality Group. 1987. *London women in the 1980s*. London: Women's Equality Group, London Strategic Policy Unit.

Women and Geography Study Group. 1984. *Geography and Gender*. London: Hutchinson.

2 Women's Workplaces: The Impact of Technological Change on Working-class Women in the Home and in the Workplace in Nineteenth-Century Montreal

BETTINA BRADBURY

Women and men are seldom affected by fundamental economic or technological transformations in the same way. Historically women's responsibility for reproduction and domestic labour as well as the particular nature of their relationship to the labour market have meant that their place in the city, and their links to the wider economy, have been gender specific. The aim of this essay is to examine how the technological changes generally associated with the emergence of industrial capitalism affected women's workplaces, in particular the home and the more formal labour market of Montreal, Quebec, during the second half of the nineteenth century. More precisely it aims to identify those periods in women's lives when they were most likely to be exposed to changing technologies and the places where such contact occurred. Labour and working-class historians have underlined the unequal impact of technological change between and within trades as well as the continued importance of hand labour throughout the industrial revolution. Here the goal is to add to our understanding of the inequality and irregularity of the impact of technological change by also considering its impact on the home. I do not intend to paint a detailed picture of changes in household technology. Rather my goal is to paint a very broad and general picture of technological changes and their impact on women by using Montreal as an example, and by drawing upon analysis of the lives of women living in Sainte Anne and Saint Jacques wards in the second half of the nineteenth century.[1]

Time and space are the lenses through which the impact of technological change on women will be assessed: the former, in terms of when in a woman's life cycle, the latter, in terms of city location and type of workplace. The transformations associated with the industrial revolution, it will be argued, affected women indirectly more often than directly. While advances in the provision of water and, later, gas and electricity had the potential to revolutionize all women's domestic labour, in this period such fundamental improvements in home conditions were mostly accessible to middle-class and more wealthy families. The labour of working-class women in the home was only slightly modified over this period. Most women wage earners continued to work in domestic service or to do homework for the sewing and shoemaking industries. Within factories their employment was most often concentrated in those parts of the labour process where hand labour or minute, repetitive tasks were intensified by transformations at other stages of the production process. While the emergence of industry did open up new jobs to women, it opened up fewer than for men. And, like unskilled men but in more dramatic numbers, women found themselves in jobs that were viewed as requiring little skill, and they were remunerated accordingly.

I will begin by sketching a brief portrait of the changes that were transforming the workshops and factories of Montreal between the 1850s and 1890s. I then will give a short outline of the major life cycle stages through which Montreal women passed during this period, identifying the relative importance of domestic and waged labour at each stage. This outline will serve as an introduction to the following two major sections of the chapter, the first of which examines technological changes in the workplace and their impact on the nature and content of women's waged work; it also briefly discusses three of the most important types of jobs held by Montreal women: domestic service in the homes of others, sewing, usually at home, and factory labour. The second major section sketches the impact of technological changes on working-class women's daily labour in the home.

THE GROWTH OF INDUSTRY IN MONTREAL

The nineteenth century was Canada's period of industrial revolution. Montreal became the country's first and major industrial city. While historians may debate the exact timing of this industrialization process, none would disagree that by the 1860s or 1870s fundamental transformations had occurred in the mercantile city's economy. By this time production was occurring not only in artisanal workshops but increasingly in large factories employing over 100 workers and using complex steam- and water-driven machinery. In 1881 over 32,000 of the city's population of 140,247 were reported to be working in more than 1,200 manufacturing establishments throughout the city (Census 1881). The magnitude of the changes that occurred is clear in the census returns for the period; they were even more dramatic for those who were caught up in the major transformations of their personal lives that rapid economic change engendered. Visitor S.P. Day expressed surprise in the mid-1860s at the variety of the city's manufacturing resources: "For miles along the banks of the Lachine Canal could be observed factories clustered together from which the hum of industry constantly went forth, passing over all together the busy malls and temples of commerce within the city itself ... Within the past five or six years, or even a briefer period, a variety of manufacturing resources have been developed" (Day 1864, 179). The machinery employed in some of these factories was also viewed with awe by contemporaries. A reporter for the *Montreal Gazette*, writing in 1864, dwelled at great length on the technological advances and the machinery in his description of the Canada Eagle Works on Saint Joseph Street: "[The machine shop was] filled with lathes, drilling machines, planes etc. One of these lathes is some thirty feet long, and its weight enormous. There are several smaller lathes used for 'finishing,' and six planing machines. All the machines are driven by steam power, and it is exceedingly interesting to witness their operation" (15 July 1864, 4). The harnessing of water power, the use of steam engines, and the application of new technology accelerated the changes in the organization of production that were already underway and, in turn, promoted new divisions of labour. Some artisans found their craft transformed as their skills were downgraded or even eliminated. Men, women, and children were drawn into the growing proletariat that made up the work force for expanding industry. They were not, however, drawn in at the same pace or on the same terms. The transformations that technological and economic change wrought were not the same for men and for women, nor for people in different class positions in the city. Nor was technological change limited to the workplace.

WOMEN'S LIFE CYCLES

Not only was women's experience of basic economic transformations and technological changes different from men's, but the impact of such changes also varied with a woman's class position, her age, and the particular stage of her individual and family life cycle. It seems important to identify those moments in a woman's life when innovations in the workplace would have been most influential but equally essential to examine those times when changes in the urban infrastructure and non-work-based technological change would have had even greater impact. We first need to sketch out the contours of women's life cycles in this century, then, to follow them from birth to death, conceptualizing their relationship to broader changes at each stage.

Few technological or scientific inventions improved survival chances for children under one year of age over this period. At birth a baby girl's chance of survival was only slightly better than a boy's. Overall up to one in three or four babies born would die before the age of one (Carpenter 1867; Olson, Thornton, and Thach 1989). For the poorer fractions of the working class, and in French-Canadian families in particular, infant and child mortality took its greatest toll; the role of different child-feeding and -weaning practices in infant mortality still needs to be examined. Certainly, had scientific and entrepreneurial effort been expended on ridding the milk sold in the city of its many impurities and diseases and on improving the water supply, the life chances of young girls and boys alike would have improved dramatically. Science was not harnessed to this end, however, and little change occurred until the twentieth century (Copp 1974). In the 1880s there was some decline in mortality rates because inoculations appear to have controlled the death rate from smallpox, but tuberculosis remained a major killer of women aged from fifteen to forty (Bradbury 1984a).

Most girls who survived the first year of life would go to school for a few years between the ages of seven and twelve, and the proportion of those who did so increased steadily as the century progressed. Girls tended to stay in school slightly longer than their brothers, who left to find the better-paying jobs that males could get and that provided essential money for the family economy. School attendance, for working-class girls in particular, was irregular, punctuated by periods of staying at home to help look after younger brothers and sisters, to do household chores when the mother was sick, or, in the poorest families, simply because they had no shoes for walking to school (Bullen 1986). Nothing that girls learned at school would lead them to expect a career or train them for skilled work in the marketplace. Montreal's school curriculums aimed to prepare girls to fulfill their expected future roles as wife and mother in ways deemed appropriate to their class position (Malouin 1983). Few girls or boys remained in school after the age of 14, and, among those aged 15 to 19, growing numbers reported having a job in this era as the possibilities of finding waged labour expanded.

The expansion of waged labour that accompanied the growth of industry in Montreal offered more work opportunities to boys and girls alike, but there were always more boys than girls who reported having a formal job. Actual percentages varied across the city according to the structure of industry in different neighbourhoods. Thus in

1881, 30 per cent of the girls aged 15 to 19 who lived in Sainte Anne, the western Montreal ward surrounding the Lachine Canal where heavy industry predominated, reported having a job, compared to 75 per cent of the boys. On the eastern side of town in Saint Jacques, where there was little heavy industry and where the sewing industry provided women with the chance to sew at home with their daughters, 38 per cent of teenaged girls reported a job, compared to only 50 per cent of boys. The structures and possibilities of local labour markets within the larger city were clearly crucial in determining what kinds of waged work sons and daughters would find. Unless they left home, their choices would be constrained by family needs and the place where the family resided.

Women's formal waged labour in these two wards usually began at around age 14 or 15 and ceased when they married, generally between the ages of 23 and 25. A minority of married women (2 to 6 per cent) reported having a job. Most, however, were totally occupied with the shopping, cleaning, cooking, childbearing, and child-caring that constituted a full-time and physically demanding job in the nineteenth century. Later in life, as widows, or when husbands became sick, some women would again find themselves seeking a paid job. In broadest terms, then, girls' work began as domestic helpers around the home at an early age. Their mothers' need for assistance with the multitude of household tasks that kept families going often kept them out of school and subsequently interrupted their wage-earning activities. As older daughters at home, they would seek waged labour, but less frequently and less regularly than their brothers. Then, as married women, most retired from the formal labour market but complemented their husbands' and other family members' wages in a myriad of different ways (Bradbury 1984b). If other family wage earners fell sick, married women might seek formal work; if the husband died and there were young children to support, widows would turn again to the more formal labour market (Bradbury 1989). Thus when we consider women's waged labour during this period, it is important to recall that it was largely, though never exclusively, performed by women prior to marriage or by those who never married, were deserted, or became widows. Domestic labour could be performed by all women at all stages of their life cycle, but it remained the primary responsibility of working-class wives and widows.

WOMEN'S WAGED LABOUR AND TECHNOLOGICAL CHANGE

How much was women's paid work influenced by the changes that accompanied the emergence of industry in Montreal? The industrial revolution involved much more than just technological change. Preceding and accompanying the use of specific mechanical innovations were reorganizations of the labour process that created new and usually less-skilled tasks. Not all industrial workers would be confronted with machines, although their particular work might be created as a result of developments in machine technology. The growth of industry increased the number of different jobs available in the labour market, as reflected in the growing number of job titles reported by teenaged boys over this period; the same is not true, however, for teenaged girls. Throughout the nineteenth century employment as a domestic remained the most often reported female occupation. Industrial work did offer an alternative to the live-in, dependent status that restricted the lives of generations of domestics, but it did not offer anything like the variety of jobs that were available to boys and men. And the types of work available varied with the economic structure of the neighbourhood. For example in the early industrial city, few workers could afford to live farther than walking distance from their work. In the Saint Jacques ward over 90 per cent of teenaged girls who reported a job in 1871 were seamstresses. Even in Sainte Anne, with its much more complex industrial structure in which male jobs predominated, over half of the working girls were seamstresses, and their places of work were probably the clothes-producing factories of the area rather than the home. The other girls doing industrial work in both of these wards were concentrated in specific types of factory work, as typecasters, tobacco workers, and in shoe, textile, or shirt and collar factories.

Factory Work

Work in factories, away from their parents, appealed to some girls because of the freedom it appeared to offer, the chance to earn money in their own names, and, in some cases, to work at jobs similar to men's; yet most appear to have turned their pay packets over to their parents. And, usually, the sexual division of labour that emerged in different factories meant that women were assigned to the subsidiary, labour-intensive jobs created as a by-product of technological change. The testimonies of employers and workers alike at the hearings of the Royal Commission on the relations of labour and capital held in Montreal in 1888, along with other contemporary evidence, suggest that women invariably performed tasks that were learned more quickly than those done by boys or young men and were remunerated at about half of the male wage. In shoemaking women were seldom taught the trade properly; they got stuck on one kind of work (*Royal Commission*

1889, 439). In one stationery, bookbinding, and printing establishment the "girls" were "employed feeding machines, sewing and folding blank account books." Boys were bonded, but there were no indentures for girls (*Royal Commission* 1889, 247). A paper box manufacturer explained that any "ordinary intelligent girl" could become adept at working in his factory in three to four months. He had a machine for making paper bags, but most girls seem to have been employed on the "great many bags of different kinds still made by hand" (*Royal Commission* 1889, 325). And, in a Montreal type-casting foundry, women were employed to hand-polish the type in order to remove the irregular bits that remained after the moulding process (*The Montreal Gazette*, 19 July 1864, 4).

One exception to this type of female employment, apparently, was in the printing trade, where women compositors did work similar to that of men in some Montreal shops and received only slightly less pay, compared to the almost standard 50 per cent less that they received in most workplaces. Furthermore one 1878 publication bears the imprint, "Montreal Women's Printing Office" (Agnes Harcourt 1878), which suggests, perhaps, that some women actually owned and operated such a workplace. In some cotton mills, too, women and men did similar jobs and apparently earned similar wages during the nineteenth and early twentieth century (Ferland 1987; Brandt 1981). In most industrial work, however, women earned roughly half of the equivalent male wage, although their ghettoization in specific limited labour processes in the majority of trades makes comparison difficult.

The Sewing Industry

Desire for independence conflicted with the rigid time requirements and discipline imposed by nineteenth-century capitalists, who were determined to shape a compliant work force. Sewing at home offered married women the chance to combine paid labour and domestic work or to supervise teenaged daughters and avoid the rigid hours and rules of factory labour. Dressmaking, like domestic service, had provided work for women in preindustrial Montreal and continued to do so throughout the nineteenth century. By the 1820s this trade was changing, as mistress dressmakers hired growing numbers of apprentices, stopped housing them while they were learning the trade, and, in the larger shops, stopped teaching them the whole trade (Poutanen 1987). There also would have been numerous Montreal women who worked as dressmakers without ever advertising their services in any newspaper and without formally engaging apprentices to help. Women sewed for their families, and some would have extended their clientele within the neighbourhood. In the 1860s clothing production was Montreal's third most important industry in terms of numbers hired. By 1881 it was the city's leading employment sector and offered a wide array of work in different kinds of workplaces. Some women still sewed in small workshops and in their homes, producing the complicated women's fashions of the period for the city's elite. Indeed women's dresses and outer clothing continued to be produced either at home or by small-scale seamstresses well into the twentieth century; but an increasing proportion of the overall production of clothing was organized by industrial capitalists, and parts of the process were taking place in large factories. Production of men's clothing – suits, overcoats, pants, jackets, and shirts, as well as underclothing, bathrobes, nightshirts, suspenders, and belts – was increasingly based in factories. So, too, was the production of those women's garments susceptible to mass production, especially bodices and underclothing, and some children's clothing (Payette-Daoust 1987).

The various needle trades involved a complex balance of factory and homework, of machine and hand labour. In 1874 one Montreal employer described his factory as having "a 15-horsepower engine running three machines having 50 needles each, and a knife which cuts the cloth by steam, so that four cutters will do the work of from twelve to fifteen." The same employer had no idea how many hands he employed: in his factory were 70 to 100 workers who prepared the work to go out, but he thought there were probably a further 700 to 1,000 outside workers, "women who live in their own homes" and who, according to him, "sit down when their breakfast, dinner and supper is over and make a garment" (Commons 1874, 23, 36).

Women dressmakers sewed by hand and by machine. Once sewing machines were available, many women bought them to save time and to increase their productivity. Since pay was calculated by the piece, it was crucial to get as much done as possible. Purchasing a machine or putting all members of the family to work helped. In this situation women's relationship to new technology was very different from that of wage labourers or artisans in a factory, where the capitalist invested in the cost of new machinery, which workers then learned to use. These women themselves invested in the sewing machines in the expectation that they would continue to receive sewing to do at home. Since few families could afford to pay the $50 to $80 that a sewing machine cost, many paid by installments of $3 to $5 a month (*Royal Commission* 1889, 603–4); for families in which a labourer head earned only $6 a week, this was a formidable commitment of money. As long as they kept earning, all went well, but nonpayment for any reason, including illness or even the death of the wage earner or his wife, meant repossession of the machine with no compensation. Changes in the payment received for sewing could mean that such investments were never recuperated. For instance one shoe manufacturer reported having introduced a buttonhole machine into his factory. He stopped using it, however, when it became cheaper for him to have work done outside. Consequently girls who had bought these machines so they could work at home were making a poor living, because the price per 100 buttonholes dropped from 60 to 16 cents. "It paid so well," he explained, "that everybody went into it; and now they are doing it for almost nothing" (*Royal Commission* 1889, 439).

Such homework in the sewing and shoemaking industries drew on the skills that many women already had, skills that were not in short supply. Piecework at minimal wages was their initiation to waged labour, which was performed in the home and thus retained elements of older patterns of women's work; yet its proliferation in mid- to late-nineteenth-century Montreal resulted from the advanced division of labour and the changing technology of factory production. Changes in the machinery used for cutting, in particular, multiplied the amount of subsequent hand labour and homework. With the exception of the sewing machine, however, the women were not using the new technology. Most were not even working in the same physical space as the machinery. Precisely because it could be performed within the home, sewing was one of the few women's jobs undertaken by married women as well as daughters and boarders. It was not uncommon in Saint Jacques Ward, in particular, to find mothers and three to five daughters all working as seamstresses. Working together, at their own pace, they could supervise younger children, prepare meals for other family wage earners, and avoid the split between home and work that earning wages outside the home involved. Their homes, as a result, offered no refuge from the work world, no parallel to the middle-class homes emerging in this period as a "haven from a heartless world."

Domestics

While sewing and factory work employed the majority of girls in working-class wards like Sainte Anne and Saint Jacques, in the city as a whole domestic service was the leading employment for Montreal women, just as it had been in preindustrial times and would be throughout the nineteenth century. Historian Claudette Lacelle argues that the only major change in this period appears to have been that "polite society had acquired the habit of eating far later" in the day than previously in the century, so that a domestic's working day continued later. Her evidence suggests that toward the end of the

century, new gadgets complicated rather than relieved domestics' tasks
(at least if cartoons of the period can be believed). In one contemporary
cartoon, the servant girl complains to the mistress that she is giving
notice, "for the likes o' this I never did see, nor will I stand. 'Ere's
Miss Amelier a poking her glass thing-a-bob inter my mince pies to
try their tempers, and well I knows as it 'tries mine, a lettin' down the
'eats and coolin' the hoven" (Lacelle 1987, 102). Yet their work may
well have been lighter than it would have been had they remained at
home helping their mothers, because it was in such wealthier homes
that inside plumbing, piped-in gas, and electricity were first used.
Electricity, in particular, eliminated much of the dirt that had characterized previous forms of lighting and heating, thereby reducing dramatically such horrific tasks as spring cleaning (Strasser 1982). No
doubt mistresses found plenty of other tasks to replace those that
shrank! And older ways of lighting and heating remained important:
oil lamps continued to be used even in urban homes well into the
twentieth century, and only a minority of homes installed electricity
before that. The vast array of "Lamps, Gas and Electric Fixtures"
displayed in the 1901 Eaton's catalogue proves the coexistence of all
these forms of lighting. The available goods advertised in the sections
on "Graniteware, Tinware, Woodenware" and "Household Furnish-
ings" underline the continuities in household implements and the lack
of practical innovation (T. Eaton Co. 1901).

Lower wages in virtually all sectors of the economy and all types of
work made it likely that most families would send sons rather than
daughters to work, when they had a choice, and that daughters would
be the ones retained at home to perform domestic labour. Low wages
for women were justified, often even rationalized, because it was
assumed that all women were secondary wage earners. Certainly many
were; young women simply could not afford to set up an independent
household, and so they did remain at home while earning wages. Over
this period girls and boys alike stayed longer with their parents, and
fewer and fewer spent time as an apprentice, servant, or boarder in
the homes of others. This new pattern in which youth remained at
home contributing to the family economy brought working-class families with wage-earning children a period of relative comfort. Teenaged
girls and young women living with their parents experienced intermit-
tent periods of waged labour, punctuated by times when household
chores were their major occupation. This served as a fitting apprenticeship for their future role as spouses. The women who did not marry,
or who became widows or were deserted by their husbands, would
fully bear the burden of women's low wages.

TECHNOLOGY AND WOMEN'S WORK
IN THE HOME

Most young women in nineteenth-century Montreal married, and the majority stopped formal waged labour immediately or shortly thereafter. As wives, then mothers, their major task was the management of the wages of others and the domestic labour that transformed those wages into sustenance, clothing, and a reasonable standard of living. In comparison with the changes that occurred in Montreal's factories and workshops, technological transformations that modified women's work in the home during this era were minimal.

By the end of this period the technology existed to install running water, gas and electric lighting, and relatively hygienic toilet facilities in houses, as well as to provide more efficient cast-iron stoves for cooking. The streets were rendered somewhat safer, for men and women alike, by the installation of gas lighting in the 1840s and 1850s. There were few domestic gas customers, however. High prices ensured that both domestic and commercial users were drawn from the local elite (Armstrong and Nelles 1986, 13–14). As early as 1878 the first public display of electrical lighting in Canada was engineered by J.A.I. Craig, a Francophone Montrealer. He set up his own small company and was joined over the next few years by a variety of other firms, many under license from American manufacturers and all competing with each other. Again, however, this new motor force, and the technologies that accompanied it, were used primarily for street lighting and secondarily for industry. As late as 1932 only 13 of 100 possible consumers in Quebec had residential electricity, and they used only 3.5 per cent of total consumption (Armstrong and Nelles 1986, 299). Not until well into the twentieth century were gas and electricity installed in most working-class homes, where eventually they lightened the necessary daily tasks of cooking and washing and eliminated much of the dirt associated with coal, wood, oil, and other older fuels, while creating new dependencies on the companies supplying the energy (Strong-Boag 1985; Strasser 1982; Cowan 1983).

Changes in the provision of water were of more benefit to nineteenth-century working-class housewives than were electricity or gas. In the 1830s Montreal was said to be better supplied with water than any other city on the North American continent, with the exception of Philadelphia (*Report of the Water Committee* 1854, 36–7; quoted in Armstrong and Nelles 1986, 15). Water sellers, though, still went from house to house. In April 1845 the waterworks came under municipal control, but temporary shortages and the use of puncheons continued. "The lack of water to wash with makes me feel the want of it," wrote one Montrealer in 1868. By the 1870s and 1880s the City supplied most Montreal dwellings with water, and all citizens paid a water tax. Usually, however, there was only one tap per household, normally in the kitchen. The implications of this inconvenience for a woman's daily work will be examined below; but since no detailed examination of the timing of these kinds of changes or of their impact on women's work in the home in Canadian cities exists, the following section simply sketches an outline of women's domestic labour in working-class Montreal homes during this period and identifies where technological change had some impact, focusing on food preparation, cooking, and washing (but see Strong-Boag 1985; 1986; 1988, 113–38; Luxton 1980; Cowan 1983; Strasser 1982).

Advances in fuel provision, running water, and toilet facilities would touch only a small fraction of the working class: the wives of skilled workers with steady, well-paid jobs. For most working-class wives, and particularly for those married to labourers or to other low-paid workers, domestic labour remained heavy work carried out in very unhealthy surroundings and was hardly changed by technological advances. The transformation of purchased or even home-produced food into edible meals was a much heavier and lengthier task than it is today. The fact that much of the food available for sale in Montreal was adulterated or poorly processed imposed extra preparation time on already busy housewives. Flour, full of impurities, had to be sieved several times, oatmeal picked over for inedible extras. Coloured confectionery was said to need careful inspection. Tea, coffee, sugar, mustard, and marinades were frequently contaminated. Tea was "diluted with stalks and teadust," and coffee was "largely adulterated with chickory, peas, roasted corn and roasted, damaged wheat" (Montreal 1879, *Annual report upon the sanitary state*, 19; *Royal Commission* 1889, 664–6). Even most unadulterated food required work to turn it into edible meals. Chickens were usually bought unplucked, fish unscaled; some wives bought bread ready-made, others made it themselves (Strasser 1982, 29). The less expensive cuts of meat, all that the poorer working-class families could afford, had to be cooked for hours before they became edible. Little wonder working-class wives were said to buy bread rather than bake it and to feed their labouring husbands on dinners of bread and cheese (Anon. 1848, 112).

Cast-iron stoves, a major advance over both the danger and the discomfort of eighteenth-century fireplace cooking, were readily available in Montreal by mid-century and were produced by city founders at lower and lower prices. Wood or coal, however, still had to be carried inside, and such stoves required constant feeding to keep them going. While men may have chopped the wood, lack of storage space within houses meant that carrying the wood or coal, often up one or two flights of stairs, usually fell to the wife or children. Furthermore any water to be heated either for cooking or washing had to be carried from the tap to the stove. Though a source of warmth in winter, such heating must have proved nearly intolerable on hot, humid Montreal summer days.

A stove, argued one English immigrant labourer in 1888, was "the one really essential piece of furniture for a labourer's family." New ranges were said to cost $20 in Montreal in the 1870s, but by the late 1880s mass production had reduced the cost to around $10, a price that was apparently fixed at that time by the stove producers (*Royal Commission* 1889, 87; Commons 1888, 370). The wives of most unskilled workers would probably have cooked on a second-hand stove, for these could be had for around $2.50 (Cross 1969, 202). In the poorest families "an old cracked stove, some chairs, a couple of cups and a tin plate" made up the cooking and eating area. A "narrow shelf against the wall" served as a table (*Montreal Daily Star*, 24 December 1883, 4). Women sharing their housing with other families might "club together, using one stove." A Montreal doctor, Dougless Decrow, testified in 1888 that in the houses of day labourers, where two families might share a house of three or four rooms, each family would "have one room for a sleeping room and use the kitchen for a dining room – the kitchen and stove in common with others" (*Royal Commission* 1889, 606).

In winter a stove was a place of congregation, often the only place to keep warm in quickly constructed, poorly built and poorly maintained houses during the long northern winter. Weary workers no doubt competed for space, not only with each other, but with the wet washing that their wives draped around the stove in an attempt to get it dry. City officials complained that this practice was so "widespread among the lower classes" as to constitute one of the major reasons for the very high incidence of fires in the city (Lapointe-Roy 1972, 30).

For housewives, drying the clothes represented the final hurdle in doing the laundry, one of the most hated aspects of nineteenth-century domestic labour. During this period the task of getting the water to do the washing became somewhat easier. Remembering the 1830s, Sarah Kurczyn describes her mother using water from barrels that had been filled from the river. Not satisfied with the supply of barrelled water available for their laundry, her mother took the clothes to the river at Point Saint Charles every Monday, finding a place where there was a green spot and "where they could be bleached." Looking back she remembered that the "preparation for this event caused great excitement in the kitchen. Tubs, wood, pots, and a well filled picnic basket had to be provided and were taken in a large square cart." Such excursions could provide moments of fun for children and an occasion for women to do their domestic tasks together, while exchanging gossip, information, or complaints. On the other hand, these chores should not be romanticized; washing, rinsing, and wringing out clothes in a river is hard, back-breaking work. Not all families could afford a cart to get to the river, and such excursions were out of the question in winter.

The backyard wells and washrooms that existed in older working-class areas like Pointe Saint Charles in the 1860s caused less upheaval than did family trips to the river, but much of the heavy work remained. Dirty clothes had to be carried downstairs or outside to be washed. It is unlikely that such washrooms were equipped with a stove, so the washing was done either with freezing cold well water or with water that was carried up to the house, heated, then brought down again. Heavy, wet clothes had to be hauled back up again by those living in upstairs apartments and were hung on a clothesline or left to dry in the house.

The piping of water to most Montreal dwellings in the 1860s and 1870s eliminated the need to carry water in from outside. Yet the fact that many residences had only one cold-water tap meant that water for cooking, washing, and cleaning still had to be carried from the tap to the stove for heating or from the tap to sinks located elsewhere, and that containers had to be carried from the tap to wherever washing and personal bathing took place. Nor was the water cost-free. All Montreal citizens paid for the water provided by the City with a water tax, which was levied on occupants, not owners, of buildings. If one could not pay this tax, the water supply was cut off. When the depression of 1873 hit, Montreal's poorer families found it impossible to pay for water, especially since tax collection occurred just before the winter, when, as the mayor acknowledged, "the attention of citizens of MODERATE means is being directed to provisions for our long and inclement winter. At the time when the cheapness of fuel renders it important for the working man to be able to lay in a stock of it to outlast the snow, he is called upon to pay in advance for the whole of the water he will consume during the year. It is easy to perceive what a strain this may place upon the resources of men who live upon daily wages, especially if from any casualty, to which they are all liable, they should happen to have been out of employment for any period of the year." (Montreal 1886, Mayor's inaugural address, 8). After five years of prolonged economic crisis, during which revenues from water taxes dropped dramatically, the city's councillors resolved to provide water as cheaply as possible, even for free, so that "poor people may have the means of attending to their personal cleanliness as well as that of their houses" (Montreal 1886, *Annual report upon the sanitary state*, 15). As late as the end of the 1880s, when the depression had pretty well lifted, nearly 800 Montreal households were reported to have had their water taxes remitted, reduced, or delayed – 43 per cent for poverty, the others for illness or for "sanitary reasons" (Montreal 1889, *Annual report upon the sanitary state*, 12). Throughout the period some of the wives struggling to balance budgets and to keep their families and households clean made trips to City Hall to protest rises in water rates or the turning off of their water supply (*Royal Commission* 1889, 262).

Horrified observers who found houses "besmeared with filth" or "overlaid with dust" (one mother confessed that "being without a supply of water, she had not been able to wash her children for ten years") witnessed one extreme in the efforts of women to deal with the tasks of washing and cleaning in homes where the amenities, technology, and infrastructure were minimal. At the other extreme were those whose "spotless cleanliness" so impressed one Catholic priest (Thébaud, 233; quoted in Cross 1969, 206). Women who had a husband with a steady and reasonable salary, or several children also earning wages, could afford to put aside part of that income to pay others to do their washing; this provided employment for poorer wives and widows in the neighbourhood, for the laundries run by Chinese immigrants (which were beginning to flourish in Montreal during this period), and for capitalists, whose steam laundries utilized the latest machinery for washing and who employed up to one hundred women to wash, iron, and fold (Lortie 1904; rpt. in Savard 1968, 106; Helly 1984; *Royal Commission* 1889, 688–90).

CONCLUSION

Steam laundries, commercial bakeries with modern equipment, and clothing factories all represented investment by industrialists in enterprises and technology that produced a good or service that had been traditionally done by women in the home. During the nineteenth century such technology, with the exception of the sewing machine, was not available in domestic form. Women's work in the home had been lightened somewhat by the availability of running water, cast-iron stoves, some improvements in lighting fuels, and by the fact that bread, for those who could afford it, could be bought relatively cheaply; others paid to have clothing washed, dried, and folded. For most working-class women, however, it was not until well into the twentieth century – when the home became vital as a market for the consumption of new commodities (like dryers and electric washing machines) and when electricity and indoor plumbing were more widespread – that technology truly began to transform their domestic labour. And those changes constituted a mixed blessing, lightening housework rather than reducing it, raising society's standards, and making domestic tasks seem less and less like real work.

In the workplace women continued to occupy particular segments of the labour market and to perform specific tasks within industry. As changing divisions of labour opened up new tasks, which were invariably viewed as unskilled, there were some changes in the types of industrial work women performed. Domestic service and sewing remained female employment ghettos and drew increasingly on immigrants, who were seen as the least likely to contest the low wages and the poor work conditions that did not change. In the twentieth century the explosion of clerical work in industry, services, and government alike opened up a whole new realm of female labour – office work – creating a vast new female work ghetto. As the twentieth century advanced, economic, demographic, and ideological changes combined to shift the timing of those moments in the life cycle when women would work both for wages and at housework. Longer schooling, shorter childbearing periods, and the costs of living in a consumer society meant that the working mother replaced the working daughter as the second wage earner in working-class families, and she became the second worker among the middle class. The lack of interest that nineteenth-century capitalists had shown in promoting technological innovation in the home was replaced by the frenzied production and promotion of appliances, all promising to render housework easier or faster for women combining waged and domestic labour.

NOTES

This paper was originally given as a talk at Concordia University several days before the Women, Work, and Place conference took place at McGill. John Bradbury, my late husband, encouraged me at that time to write it up and to submit it for consideration for this volume. It seems appropriate in an article that, in part, deals with balancing formal waged labour and domestic labour to acknowledge all the dishwashing, clotheswashing, ironing, and other domestic tasks that John did over the years and that helped me as I combined mothering with work on the thesis, articles, and the book. I have sorely missed this help since his death. Few of the women whose lives I try to describe in this chapter would have been able to say the same of their husbands!

1 The data upon which my discussion of women's life cycles and occupations is based are derived from random samples of the families enumerated in the Sainte Anne and Saint Jacques wards of Montreal, as collected in the manuscript censuses of 1861, 1871, and 1881 (Ottawa). The 10 per cent samples resulted in a total sample of 10,967 people living in 2,278 families. For further details about this methodology and other aspects of the family economy, see Bradbury 1984a.

REFERENCES

Anon. 1848. *The Skillful Housewife's Guide: A Book of Domestic Cookery Compiled from the Best Authors.* Montreal: Armour and Ramsay.

Armstrong, Christopher and H.V. Nelles. 1986. *Monopoly's Moment: The Organization and Regulation of Canadian Utilities, 1830–1930.* Philadelphia: Temple University Press.

Bradbury, Bettina. 1984a. The Working Class Family Economy: Montreal, 1861–1881. PhD diss., Concordia University.

– 1984b. "Pigs, cows and boarders: non-wage forms of survival among Montreal families, 1861–91." *Labour/Le Travail* 14: 9–46.

– 1989. "Surviving as a widow in 19th-century Montreal." *Urban History Review/Revue d'histoire urbaine* 17. 3: 148–60.

– 1993. *Working Families: Age, Gender and Daily Survival in Industrializing Montreal.* Toronto: McClelland and Stewart.

Brandt, Gail Cuthbert. 1981."'Weaving it together': life cycle and the industrial experience of female cotton workers in Quebec, 1910–1950." *Labour/Le Travail* 7: 113–26.

Bullen, John. 1986. "Hidden workers: child labour and the family economy in late nineteenth-century urban Ontario." *Labour/Le Travail* 18: 163–87.

Canada. House of Commons. 1874. *Report of the select committee on the manufacturing interests of the dominion, 1874.* Ottawa.

– House of Commons. 1888. *Report of the select committee to investigate and report upon alleged combinations in manufactures, trade and insurance in Canada,* appendix 3. Ottawa.

– *Royal Commission.* 1889. *Royal commission on the relations of labour and capital.* Quebec evidence, vol. 3. Ottawa.

Carpenter, P.P. 1867. "On the vital statistics of Montreal." *The Canadian Naturalist and Geologist* 3: 134–56.

Copp, Terry. 1974. *The Anatomy of Poverty: The Condition of the Working Class in Montreal, 1897–1929.* Toronto: McClelland and Stewart.

Cowan, Ruth Schwartz. 1983. *More Work for Mother: The Ironies of Household Technology from the Open Hearth to the Microwave.* New York: Basic Books.

Cross, Suzanne D. 1969. The Irish in Montreal, 1867–1896. Master's thesis, McGill University.

Day, S.P. 1864. *English America: or Pictures of Canadian Places and People.* London: T. Cautley Newby.

Ferland, Jacques. 1987. "Syndicalisme 'parcellaire' et syndicalisme 'collectif': une interprétation socio-technique des conflits ouvriers dans deux industries québécoises (1880–1914)." *Labour/Le Travail* 6 & 7: 49–88.

Harcourt, Agnes. 1878. *For His Sake.* Montreal: Montreal Women's Printing Office.

Helly, Denise. 1984. "Les buandiers chinois de Montréal au tournant du siècle." *Recherches sociographiques* 25. 3: 343–65.

Lacelle, Claudette. 1987. *Urban domestic servants in 19th-century Canada.* Report prepared for Minister of Supply and Services Canada. Ottawa.

Lapointe-Roy, Huguette. 1972. *Paupérisme et assistance sociale à Montréal, 1832–1865.* Master's thesis, McGill University.

Lortie, S.A. 1968. "Compositeur typographe de Québec en 1903." In *Paysans et ouvriers Québécois d'autrefois*, Pierre Savard, ed, 79–150. Quebec: Les Presses de l'Université Laval.

Luxton, Meg. 1980. *More than a Labour of Love: Three Generations of Women's Work in the Home.* Toronto: The Women's Press.

Malouin, Marie-Paule. 1983. "Les rapports entre l'école privée et l'école publique: l'académie Marie-Rose au 19e siècle." In *Maîtresses de maison, maîtresses d'école*, Nadia Fahmy-Eid and Micheline Dumont, eds, 77–92. Montreal: Boréal Express.

Montreal. 1866–1889. *Annual reports upon the accounts of the corporation of the City of Montreal* (MAR). Montreal: Diverse publishers.

Olson, Sherry, Pat Thornton, and Quoc Thuy Thach. 1989. "Dimensions sociales de la mortalité infantile à Montréal au milieu du 19e siècle." *Annales de démographie historique 1988*: 299–325.

Payette Daoust, Michele. 1987. The Montreal garment industry, 1871–1901. Master's thesis, McGill University.

Poutanen, Mary Anne. 1987. "For the benefit of the master": women and dressmaking in Montreal, 1825–1840. Master's thesis, McGill University.

Strasser, Susan. 1982. *Never Done: A History of American Housework.* New York: Pantheon Books.

Strong-Boag, Veronica. 1985. "Discovering the home: the last 150 years of domestic work in Canada." In *Women's Paid and Unpaid Work: Historical and Contemporary Perspectives*, Paula Bourne, ed, 35–60. Toronto: New Hogtown Press.

– 1986. "Keeping house in God's country: Canadian women at work in the home." In *On the Job: Confronting the Labour Process in Canada*, Craig Heron and Robert Storey, eds, 124–51. Montreal: McGill-Queen's University Press.

– 1988. *The New Day Recalled: Lives of Girls and Women in English Canada, 1919–1939.* Toronto: Copp Clark Pitman.

T. Eaton Co. Ltd. 1970. *The 1901 editions of the T. Eaton Co. Limited Catalogues for Spring & Summer, Fall & Winter.* Reproduced with an introduction by Jack Stoddart. Toronto: Musson Book Company.

3 For the Sake of the Children: Japanese/Canadian Workers/Mothers

AUDREY KOBAYASHI

intro ⇒ During the last decade of the nineteenth and the first two decades of the twentieth century a labour migration of between 20,000 and 30,000 persons from Japan to Canada took place. Approximately 5,000 of these immigrants were young women, the majority of them "picture brides" – so called because the marriages were arranged by the families, using photographs as a form of introduction. This paper examines the experiences of these Issei (first generation) women, first, by addressing the structural changes in what were publicly recognized as women's productive and reproductive roles in Japan during the Meiji period (1868–1912), and second, by discussing their transition from Japanese agrarian society to Canadian industrial society. This transi- ⇐ *intro* tion was strongly conditioned by three factors: the dominant ideology of late Meiji Japan, which advanced the economic objectives of the government; the patriarchal traditions of the agricultural village, which arose partly as a form of adjustment to national objectives and changes imposed through modernization; and the constraints that arose within a Canadian society dominated by racist ideology. Within this context to be both Japanese and Canadian, a worker and a mother created a double set of contradictions that defined the material and the emotional experiences of these remarkable women.

MEIJI PERIOD DEVELOPMENT

The Meiji government, in undertaking a highly organized and effective programme of industrialization, did not follow the classical process of capitalist development as seen a century earlier in Britain. Rather than begin with structural changes to the agrarian sector followed by the development of labour-intensive industries, Japan started its development process with the establishment of capital-intensive industries – transportation and communications, and armaments – while simultaneously strengthening the agrarian economic base from which the majority of taxes derived. Labour-intensive industries, notably textiles, became important only in the 1880s and thereafter dominated export production for some time. By the turn of the century the original government enterprises had been transferred into private control, and the pattern for Japan's peculiar form of industrial production had been set. By thus reversing the development process and by restructuring the agrarian economy *in situ*, it is argued, Japan was able to modernize quickly, to control the rate and direction of social and ideological change (maintaining stability in the face of economic hardship), and to minimize foreign debt and economic dependence on foreign powers. This programme changed the lives of the 90 per cent of the Japanese population engaged in farming by intensifying the institutional relationship between government and citizen, by transforming the economic basis of farming, and by initiating changes in everyday traditions, particularly the structure of the family.[1]

From the beginning the influence of government upon its citizens increased with the institutions it imposed: universal education, military conscription, medical services, and centralized programmes to improve farming productivity. Education and the military served both to create a work force suitable to a developing industrial nation (with 95 per cent school enrolment of children by the turn of the century), and to provide the vehicles of control with which an ideological system was imposed. Improved medical care, combined with education, decreased the death rate and, in combination with a reduction in the practice of female infanticide, contributed to the demographic transition. Improved farming methods helped to increase production in the early years. The government was also instrumental in promoting local organizations – young men's associations, old men's associations, women's groups, self-help groups – that also provided settings for the reinforcement of social values.

The dominant ideological thrust at the time was the creation of national values that conformed to a liberal economic system. Most important were the complementary notions of private ownership of property and the strengthening of the individual household through which private property was held. Maintenance of these values was deemed imperative by the state if the countryside was to continue to produce both the food and the taxes upon which the modernization programme depended. The government required deliberate and effective ideological imposition in order to control the relationship between stability and change.   Code

The farming population responded by transforming the village social structure from a so-called feudal system based on mutual hereditary obligation to one in which private property was the predominant measure of status. Farming became more and more valued as an ideal way of life as the stakes in the system escalated. Social customs such as marriage, inheritance, and the setting up of branch households became oriented toward the preservation of private property. Within the household, although human relations continued to rest on long established moral precedents, family structure had adjusted to the new economic and political context (Koyama 1962), with an emphasis upon "loyalty and filial piety" extending from personal relations to the metaphor of extended family that had helped to define the Japanese nation-state (Nolte 1983). The moral imperatives of village life were the foundation of economic development within a system that could be described as both patriarchal and paternalistic.

The agricultural population, however, was affected in ways that were not economically beneficial in the short term. Higher taxes, newly imposed costs of education, and reduction of the work force through conscription took their toll, although effects varied according to economic circumstances in different parts of the country. In many areas traditional means of livelihood were destroyed or severely cut back because the local cottage industries (such as cotton spinning and weaving) could not compete with foreign imports or mechanized production; because areas were cut off by the new system of transportation; or perhaps because formerly cooperative ventures, such as charcoal production, were no longer feasible in a money economy. Some forms of rural production flourished (for example, silk production and the manufacture of food products such as sake and soya sauce) in part because policies of the Meiji government encouraged a strong sense of entrepreneurship in the rural areas. Textile factories were originally situated throughout the countryside to draw upon the cheap female labour that was available in farming villages. This system had little tolerance, however, for variations in levels of production. Under intense economic pressure to produce, especially during the 1880s with the imposition of a deflationary monetarist policy, a crop failure could mean quite literally the end of a household's resources. Following massive crop failures in 1869, 1884, 1897, 1902, 1904, and 1910, periods of intense hardship and starvation and the

systematic and progressive creation of rural poverty occurred (Irokawa 1985). Whereas the overall wealth of the country increased
tremendously during the Meiji period, not everyone shared its benefits.

By the 1880s and 1890s an ironic contradiction had developed: as
the ideal of the noble farmer carrying on his ancestral heritage through
private property ownership increased in strength, so did the level of
tenancy, and agrarian society became polarized rather than levelled
throughout the Meiji period. Rapid growth in the population, coupled
with regional economic stress, made it increasingly difficult for the
ideal of individual property ownership to be met. Many farmers who
had felt the flush of independent prosperity at the beginning of the
period could not sustain their holdings and fell into tenancy, while a
few others amassed great fortunes. No longer, however, was Japan a
land of peasants content with their lot in life; it was a society united
in its quest for economic success and thoroughly tied to a modern
world economic system. Though the Meiji ideology was hegemonic,
this was not a naïve society but one channelled by deep currents of
discontent with the status quo, churning with potential for innovation
in turning events to advantage.

One practice that developed to counter this contradiction between
normative ideals and lived conditions was *dekasegi*, a euphemism
meaning "temporary migrant labour." In principle *dekasegi* was a
temporary solution to economic difficulties: young men and women
would be contracted to industrial enterprises in order to improve their
circumstances, so that they could return to marry and take up an
agrarian life in the village. The term is euphemistic because, for all the
stories of success in city life, the vast majority of these young people,
especially those who were not direct-line heirs, never returned to their
villages but instead became part of the growing urban industrial work
force. Their returns to the village usually were due not to the pull of
an agrarian dream but to periods of unemployment in the urban areas.
Those who managed to return to the village permanently, mainly first
sons or the limited number of others who inherited property, might
have spent long periods of their lives in industrial labour on a seasonal
basis.

These changes in the work patterns of the rural population occurred
as the Meiji government effected the second stage of its policy with
the development of labour-intensive, mainly textile, industries in the
late 1880s. Proletarianization escalated a decade later, after the Sino-
Japanese War, with the sale, at bargain prices, of the heretofore
government-run strategic industries into private hands. (The result of
such sales was the formation of integrated, multinational concerns,

such as Mitsui and Mitsubishi, known as the _zaibatsu_). By the turn of
the century patterns of modernization that would strongly influence
Japanese society well into the future had been set; farming continued
to occupy an ideal position and to provide the stable, conservative
backdrop for an increasingly industrial society made up of an educated
but also deeply traditional (in the new sense of that term) work force.
Changes in social relations and family structure had been effected to
create a society well prepared for the plunge into modernity.

WOMEN IN MEIJI JAPAN

Two major types of change took place for women at this time in
Japanese history: their reproductive roles altered as greater emphasis
was placed upon the continuation of the household; and their productive roles altered as they shifted from being an integral part of the
rural economy of agriculture and cottage industry to being members
of a heavily exploited industrial work force. The value of the female
increased because she was needed in the role of mother to increase the
population and to provide the domestic stability in which that population could be nurtured and educated. Her value as a waged worker
also increased, because, for the first time, her worth could be measured
in real wages. Her increased value led not to emancipation, however,
but to a strengthening of patriarchal relations, in both senses of that
term: in the systematic subordination of women to men throughout
society, and in the dominance of the father or male household head.

The institution of marriage was modified to conform to the values
of liberal economics that marked the Meiji period. Heretofore, in the
absence of private property ownership as a basis for social organization, there had been little emphasis placed upon reproducing the
household through a male heir. The size of population had remained
relatively stable and strongly tied, by both custom and legal obligations, to the land (Smith 1970). Women's work was important in the
fields or in cottage industries but could be combined easily with the
duties of motherhood. Personal criteria, such as the feelings of the
couple involved, were more important as the basis for marriage; only
in the upper classes were marriages arranged with a view to economic
and political ends. Increasingly throughout the Meiji period, however,
the marriage contract became an important means of securing the
future of the household and protecting its economic and social
interests. For the lower classes, following the lead of the upper classes,
arranged marriages became the norm in what Befu (1971, 52) calls
the "samuraization" of the farming class. A marriage became a
practical liaison, prospective in-laws more demanding of suitable

brides.[2] The divorce rate increased as women who were deemed unsuitable, especially if they did not quickly produce heirs, were sent back to their families in disgrace (Kawashima and Steiner 1960). This system was supported by the state, since the qualities of a good wife – as well as the penalties for neglecting her duty – were prescribed by law under the Meiji Civil Code (Hendry 1981; Robins-Mowry 1983).

With the expansion of the textile industry, however, the demand for good mothers was countered by the demand for good workers. By the turn of the century textiles had become the most important Japanese export commodity, and 60 to 80 per cent of the textile labour force was made up of women (Hane 1986, 144–5; Robins-Mowry 1983, 36–7; Nolte 1983, 4). Not only were industrial workers needed for expanding industry, but as economic conditions in the countryside deteriorated, beginning in the 1880s, the efforts of all members of the household were drawn upon in order to ensure not only their own futures but the viability of the ancestral property. The reciprocal relationship between agrarian and industrial work that resulted from the practice of *dekasegi* meant that exogenous marriages were delayed while young women devoted their earliest adult years to household interests. They existed under conditions of virtual slavery; arrangements were made by their fathers or grandfathers for them to undertake a contract of specified duration before returning to the village and an arranged marriage. Life in the spinning factories was hellish: they were worked mercilessly, underfed, and poorly housed. Tuberculosis and other industry-related diseases were rife, and a significant number did not live to fulfill their contracts.[3]

The state was crucial in creating this dual patriarchal structure. Nolte (1983) argues that the roles of mother and worker were complementary within a system that applied the family motif to the larger structure of state organization in which the central values of loyalty and filial piety were extended from the individual to the nation. The nature of female experience was defined, according to her dual purpose, in several ways: from laws that specified her position in the Civil Code, to laws, such as the Public Peace Police Law (1900), that prohibited her from participation in political activities; from the insertion of antifeminist rhetoric in all forms of public activity, to the support of women's associations (*fujin kai*) that served as vessels of femininity. Factory labour, Nolte argues, was an extension of obligations to the household, overshadowed only in the 1920s and 1930s with the rise of a permanent male work force in heavy industry, accompanied by reinforcement of the "wise mother" image that "accorded with the general paradigm of women's unselfish devotion and exclusion from politics" (6).

Despite this complementarity, however, a contradiction existed that, firmly rooted in class distinctions, continues in Japanese life today. That is, the reproductive role of women is associated with the luxury of the upper classes in being able to keep their women at home. It embraces an image of the nurturing mother, whose place is in the home where she bears children and prepares them for adult life: girls, as future mothers; first sons, as inheritors of the family land and the responsibility for the household; other sons, for the best possible circumstances outside the ancestral home. The mother must foster both the education and the social connections that would assure her children a place in this world. Since the 1880s, however, women, especially young, poor women, have played an industrial role antithetical to such conservative values. They have worked beside men and have made an important contribution to Japanese production; yet their work has been denied importance and disguised, which entails serious consequences both for their working conditions and social recognition. The denial occurs because the upper class image of mother/wife forms the ideal, while the image of the worker (always, for women, subordinate and oppressed) is shunned. The overwhelming objective in the lives of Japanese women, therefore, was, and is, to become mothers. If that objective has become easier to obtain in recent Japanese history, it is because of economic restructuring, not because of a change in cultural values.

The Japanese definition of "mother" was culturally constructed to conform to demands expressed within a broad context, and the life cycle was structured according to these demands (Bernstein 1976). Despite its hint of American moral condescension, Alice Bacon's study of nineteenth-century Japanese women captures this essence of their motherhood: "The Japanese mother's life is one of perfect devotion to her children; she is their willing slave. Her days are spent in caring for them, her evenings in watching over them; and she spares neither time nor trouble in doing anything for their comfort and pleasure ... The Japanese woman has so few on whom to lavish her affection, so little to live for beside her children, and no hopes in the future except through them, that it is no wonder that she devotes her life to their care and service, deeming the drudgery that custom requires of her for them the easiest of all her duties" (Bacon 1891, 101–2).

No disappointment in life was greater than to be childless: "The lot of a childless wife in Japan is a sad one. Not only is she denied the hopes and the pleasures of a mother in her children, but she is an object of pity to her friends, and well does she know that Confucius has laid down the law that a man is justified in divorcing a childless wife" (Bacon 1891, 102–3). This "feminine ideal" stems from patriarchy and is defined and reinforced through all aspects of social life, from religion to economy (Paulson 1976). The process is a dynamic one: the female gained value because of her important role as reproducer and producer within the Meiji economy; this value was generated specifically within Japan's peculiar system of capitalism. Patriarchy, however, predated capitalism and is rooted in a long history of religious and military ideals. It was not so much a product of capitalism, then, as an existing channel along which capitalism could serve its ends. Meiji patriarchy was not simply a carry-over from a traditional system, however, for it was, if anything, strengthened during the period, its structure transformed in its role as both the means to and the desired end of capitalist ideology.

Our objective, then, should be to analyze the "combination of patriarchy and capitalism" (Hartmann 1979, 1) and the effect of this combination of ideologies on women's lives. This type of analysis requires two things: one, that gender relations be seen as socially constructed, not only between men and women directly, but between women and the state, women and children, and women and women as they share the often emotional consequences of these relationships; and two, that the spatial extent of gender relations encompasses a number of scales, from home to workplace to nation, and that shifting spatial relations be taken into account. These different places, moreover, contain relationships constructed not through a separation of the spheres of home and workplace but rather as a result of the intersection between the two (Alexander and Taylor 1981, 370) – and with a wider definition of spheres that includes society as a whole.

Patriarchy is also much more than the oppression of women. It is bound up within a cultural system, a shared way of life, in which gender relations consist of "both the conflict and the complementary association between the sexes" (Rowbotham 1981, 366). Like the image of "farmer" that provided the ideal basis for material wealth and social status earlier in Japan, the image of "mother" represented the pinnacle of rightness and therefore, for most, the greatest source of happiness in a Japanese woman's life. Hence the life of the Japanese woman came to be ordered by a mutually reinforcing pattern of "constraint and fulfilment" (Lebra 1984). To grasp this dialectic will require more than uncovering the source of women's oppression; it will require an understanding of the tremendous emotional force with which gender relations are forged.

These observations hold especially in times of social change. Meiji Japan was a time and place of rapid social change in the wake of state intervention that hastened both ideological and economic transition and that used established cultural values to cushion the effects. Against the stress created by the resultant contradictions, people tended to cling to traditional structures in the hope that these would provide security and ease the adjustment to massive change in other areas of life. It is precisely at such times, however, that traditional structures, designed for other times and places, fail. One of the keys to theoretical understanding of cultural (and gender) relations is the analysis of that failure – and of the ways in which traditional structures are subsequently rejected, modified, replaced, or reinforced (witness, for example, the sweeping changes in marriage patterns) as "tradition" is continually redefined. Any project embarked upon either to further the objectives of or to overcome the obstructions cast before Meiji society invoked this pattern. One such example was international migration, a late Meiji development that restructured social and economic circumstances, forced a major readjustment of gender relations, and yet was undertaken with the object of preserving rural stability.

EMIGRATION TO CANADA

As already stated the Japanese people, despite the abusive constraints of a rapidly industrializing nation, were no mere pawns in the hands of a government gone wild with power. Under a variety of conditions they responded to the circumstances set before them and used their recently acquired education and skills to change their lives in many ways. Although these responses sometimes took the form of resistance to government and industrial authority, especially in the urban areas, they were more often conservative, designed to reinforce the ideals of agrarian society rather than to destroy them. One such response was international migration; no more than a form of *dekasegi* in principle, it was viewed by the migrants themselves as a means to obtain money (much more than could be earned in factories in Japan) in order to be able to return eventually to the village, to become established in the cherished life of farming (Kobayashi 1983, 1984). Although not all of the emigrants returned to Japan, and although their subsequent experiences in Canada changed their lives, the decision to emigrate must be viewed contextually, as part of the process of modernization outlined above. Labour migration to Canada can be divided into three distinct stages (Kobayashi 1986). From approximately 1880 to 1908, it was almost exclusively confined to men who travelled to Canada, worked for periods that varied from several months to several years, and then returned to Japan, some to farming, many others to industrial jobs. The majority worked in the lumber industry and fishing, and a substantial number established themselves as merchants. A case study of emigration from a village in Shiga Prefecture shows that the first male emigrants were heads of households and first sons, those who sought to improve the financial circumstances of the household in Japan (Kobayashi 1983, 1985). These individuals realized substantial increases in their village holdings, built houses, and made large donations to the religious institutions of the village, thus reinforcing their hold on the agrarian way of life so strongly advocated under Meiji ideology while using an unconventional means to do so.[4] As the goals of this most important group of village men were met, second and third sons were introduced to the labour emigration process. They undertook work in Canada only as a means to establish themselves independently. Some stayed permanently in Canada; others returned to life in Japanese cities; a few saved enough money to return to their villages, buy land, and become independent farmers.

For women this first period of labour emigration had two consequences. First, because of the enhanced earning power of the men, the agrarian ideals – including the basic concepts of patriarchy – were strengthened. It became even more important for women to play their parts in order to protect the increased economic interests of the household. Secondly those women who were left behind while husbands and sons went to Canada gained greater responsibility for the affairs of the household and learned new ways of organizing family life from across the ocean. A small number of these women accompanied their husbands to Canada for short periods and there worked in supportive positions. Children born at this time were usually returned to the village (either taken by their mothers or sent with relatives) to be cared for within the extended family. Canada was not at this time viewed as an appropriate environment for raising children and, as a result, mothers were often deprived of the pleasure of raising them at all.

The second stage of emigration occurred after the Hayashi-Lemieux "Gentlemen's" Agreement of 1908, entered into in reaction to racist riots in the streets of Vancouver when White supremacists attacked the Chinese and Japanese immigrant quarters (Adachi 1976, 63–86). Japanese immigration had peaked at 7,601 individuals in 1907–08, after a particularly severe crop failure in 1906. The agreement limited future male immigrants to 400 per year but placed no restriction on immediate family members. The practice of ready movement back and forth across the Pacific stopped, and instead there began a substantial flow of young women, some already married and perhaps accompanied by children, the majority of them picture brides. The ratio of men to women, which had been 30:1 in 1893, dropped to 5:1 by 1910 and dropped further to 2:1 by 1920 (Department of External Affairs 1920, *Annual consular report*).

Picture marriages (*shasshin kekkon*) occurred almost entirely on pragmatic grounds. The women were seldom willing partners; they were chosen because they had reached marriageable age and did not have better prospects, and because there was a shortage of young women for the men who were abroad. These young men needed both to fulfill their family obligations by producing heirs and to ease the anxieties of the Japanese consulate in Vancouver, which was determined to show its citizens in good moral light to the larger Canadian community. The claim that "the arrival of the Japanese wives ... broke up a tendency among the immigrants towards increasing immorality, and the wives were responsible for mitigating the incidence of degeneracy" (Adachi 1976, 91) is doubtful, although there was a general concern that the immigrant society would fare better if it could be seen as settled and demographically balanced rather than as a collection of itinerant bachelors. Wives would not only keep the men at home but would contribute to the household income by engaging in waged labour or, among the upper classes of Japanese immigrants, working in the family business or on the family farm.

The majority of those who emigrated to Canada at this stage returned with their families to Japan and there took up again the threads of Japanese life in some fashion. A smaller group, made up in general of households headed by second and third sons who had no obligations to carry on the family tradition in Japan, remained in Canada. By the time of the third emigration stage, during the interwar period, permanent migrants established what we now know as Japanese-Canadian society. Their concern was with raising Canadian families.

From interviews with the women who came to Canada in the first decades of the twentieth century, I have found that, generally speaking, those who returned to Japan are bitter about their experiences in Canada.[6] They regard those years as those of exile, endured only for the sake of the family. The other group, those who remained in Canada – and there are now only a few hundred of them still alive – have never forgotten the wrenching circumstances under which they were pulled from their childhood surroundings. Both groups tell poignant stories filled with lessons about how we, as social scientists, might root

our structural accounts of broad political, economic, and social change
in the everyday lives of women – women whose difficult experiences
map the changes that occur as people respond to changing conditions
of living. They have vivid memories of the hardships of the early years
in Canada and, especially for those who returned to Japan after 1945,
of the shame and misery that was caused by their uprooting and
dispossession during the 1940s. Nonetheless they show a toughness
and adaptability, a sense of fulfillment for what they have achieved in
Canada, especially with reference to their children and grandchildren.
Over and again they use the term "*kodomo no tame ni*" (for the sake
of the children) to justify their actions or to make today's hardships
seem fleeting against tomorrow's dreams.[7]

The women recall the arrangement of marriage with a sense of
fatalism. Most set out from Japan as teenagers, obeying their parents'
wishes, not knowing what they would encounter. A few recall the
sense of adventure with which they set forth, either alone or in the
company of relatives or other picture brides. Some were eager to be
away from what seemed a needlessly restrictive society; a larger
number were terrified of their unknown fates. These, however, were
passing emotions, usually dulled by a month-long journey. Overriding
all other reactions was the acceptance of the fact that their own
wishes, had they been expressed, were of little concern. It is perhaps
this sense of inevitability, of life having been beyond the control of
the individual, that accounts for the consistency in the way in which
Issei women, different and individual in many ways, structure their
emotions.

One of the best ways of "reading" this structured emotion is in the
haiku, a traditional poetic form that expresses the essential qualities
of human experience and that was written by a large number of
Japanese immigrants (Kobayashi 1980). A common theme, especially
for women, is the wrenching experience of leaving Japan:

> Parting tearfully,
> Holding a one-way ticket
> I sailed far away.
> (Yukiko; trans. Ito 1973, 248)

This is not simply an expression of individual sadness but a structured
response that draws upon hundreds of years of vernacular literary
tradition in which a woman's tears are not so much sad as poignant,
an expression of burdens that go beyond immediate suffering. They
also represent, in this sense, the fulfillment of expectation. The choice
to go or stay was not her own; at the end of the one-way ticket was

an obligation that transcended individual need and solidified her filial obligations. If, in Canada, she subsequently made a happy or "good" marriage, it was because of a combination of fortuitous relations with the chosen mate and the training she had received throughout her life to be accepting, to be happy with her lot, to adjust, and, later on, to turn to her children for her greatest sense of accomplishment.

For women disembarking in Vancouver, the most painful aspects of adaptation to a new life began. First, the shy initial encounter with a husband. Some found that the husbands matched their expectations, while others found that the husbands had sent photographs of younger or more handsome friends (Ito 1973, 188–95). Within twenty-four hours came the humiliating experience of being whisked to a drygoods shop to shed the Japanese-style kimono that was a mark of shame to those already living in Canada and to replace it with a Western-style dress over (for the first time) underwear and a corset; thence to the beauty salon to dismantle the Gibson-girl hairstyle current in Japan at the time but deemed ridiculous and overdone in turn-of-the-century Vancouver and ultraconservative Victoria. An elderly woman recalls: "I finally arrived in Victoria, BC on January 23, 1913 after almost a month of travel. Toyoki was there to meet me. I was really too exhausted and too excited to notice what manner of man I was to have for my husband. I was dressed in my newest outfit, but what a strange sight I must have been to the Canadians. I had on a blue *hakama* [long Japanese skirt] and a *hawori* [blouse length kimono]. I wore a too large pair of shoes which were considered quite elegant in Japan at that time. How I wish I could show you a snap of me dressed like that. You would have a good laugh. My hair was puffed out all around in a pompadour and secured in a knot at top" (Moriyama 1982, 8–9). The young bride was introduced to her new residence, often as not a much poorer abode than that from which she had come in Japan. If it was in the city of Vancouver it was likely to be a cramped hotel room, or part of a boarding house, usually a two-storey wooden structure with any number of rooms added on haphazardly to house perhaps five or ten families or up to thirty or forty single men. The sleeping facilities in such places consisted of wooden boxes stacked one on top of the other (Kobayashi 1992a). Alternatively the new couple may have set out for some rural part of the province, where accommodation was in a flimsy shack on the site of a cannery or lumber camp or, for the lucky ones, on a farm. One woman remembers her arrival: "It was in the middle November and pouring rain. At the Port Hammond train station, his brother-in-law came with a wagon to meet us. Bouncing up and down, we travelled a brush road. At the end of it,

I noticed what to me were strange shacks. They looked like the houses of beggars. So even in Canada, I thought, there must be beggars. Then our wagon stopped before the smallest shack of all, and I was told, 'This is your house'" (Japanese-Canadian Centennial Project 1978, 18–19). The new bride would be required to be up at dawn on the day after arrival, the day after meeting her husband for the first time, to work in a laundry or family business or perhaps to cook for a camp of labourers under conditions that were, of course, entirely foreign. She would endure the insults of a husband whose main concern in bringing her from Japan had been to provide food and clean clothing for the men under his supervision in the bush camp, and who did not want to be embarrassed by a wife who was inept at her new work. Under such conditions, one woman recalled, "My tears were silent and helpless."

WOMEN'S WORK

Substantive evidence of the working lives of Issei women is very difficult to find, partly for lack of data and partly because the data that do exist are influenced by their implicit ideological definition of woman. For example semiannual reports of the Japanese Consulate in Vancouver to the Department of External Affairs in Tokyo provide population figures by industry and by sex; and, from 1922, they also break down the population into "workers" and "dependents." These data provide an excellent source for understanding the participation of male Japanese immigrants within the British Columbia economy. The data are gathered, however, under the assumption that a "household" consists of a male head and his dependent wife and children. Although the reports do not clearly specify how the counts are taken, the very small number of women represented in the "worker" category would indicate that something is seriously amiss with this information. Records for the year 1922,[8] for example, show the following:

Table 1
Japanese Canadians in British Columbia

Category	Total
Working males	9,491
Working females	198
Dependent males	4,340
Dependent females	6,459
Total Population	18,488

Source: Japan, Dept of Foreign Affairs 1922

Table 2
Japanese-Canadian Women's Work

Category	Total Employed
Independent agriculture	4
Farm labour	4
Seamstress	24
Laundry	7
Hairdresser	37
Factory labour	11
Sales	11
Boardinghouse/restaurant	8
Railway labour	3
Education	7
Midwife	9
Massage	3
Journalist	2
Domestic	57
Student	10
No data	1
Total	198

Source: Japan, Dept of External Affairs, 1922

The small total, which shows only about 10 per cent of working-age women officially employed, contradicts all circumstantial evidence. All written and oral accounts claim that women worked in factories, in domestic service, on farms, and in family businesses. In the absence of further information it becomes necessary to speculate that the only women included in the above statistics are single women. This conclusion would correspond to the Meiji Civil Code, in which it was specified that single women could head a household (work independently), while married women could not.[9] The contradiction can be explained if we examine the social definition of women's work. Several categories – education, tailoring, hairdressing, massage, midwifery – were considered appropriate for women, especially in cases of widow- or spinsterhood, and these figures are likely accurate. The small number of independent women farmers is almost certainly made up of widows, since records show a few examples of widowed women becoming heads of households (e.g., Nakayama 1982, 108–10). The major areas of factory and domestic work are most definitely underreported, since we know both from published sources of the time and from contemporary interviews that nearly all the wives of fishermen worked in fish canneries. Archival photographs show rows of women working in the canneries, babies strapped to their backs and young children playing at their feet (Steveston Imperial Cannery Collection). Interviews indicate that domestic service was very common, but, since it would have been for

(margin annotations: "MP", "Transfer of ethics")

cash payment and because it was considered most demeaning, it was in all likelihood not often officially reported.[10] A plausible explanation for the small number listed in domestic and factory work is that there were some single women registered independently prior to marriage, whereas married working women were not registered. Women working beside their husbands in enterprises such as commerce or farming seem not to have been reported at all, since their work was considered to be in support of the household head, therefore undertaken without direct payment and not listed as official work.

Personal accounts of women's lives give a more accurate, if selective, picture of their work and of the ways in which their work was viewed. A collection of over 700 short individual autobiographies produced just after World War I (Nakayama 1921a)[11] contains fifty-two such accounts. The selected women fall into three categories: the wives of wealthy or highly respected individuals, such as businessmen, ministers, teachers, or consular officials; women who are singled out for being "brave, loyal and submissive" (Shibata et al. 1977, 35), whose noteworthy activities include volunteer services in local hospitals, strong support for husbands in their business enterprises, fortitude in supporting the family in the absence of a husband or in widowhood, and dedication to the principles of Buddhism; and women who had made notable achievements in the feminine occupations of teaching or midwifery.

Most women, however, followed their husbands to Canada and to lives of arduous toil punctuated by the birth of children, who often represented only a minor disruption in their working lives. The accounts of these women's lives all stress the need to work to support their husbands. The more fortunate worked in a family enterprise, such as a business in the Powell Street area of Vancouver, or on a farm in the Fraser or Okanagan Valleys. This privileged group played a matriarchal role and was responsible for the gentle introduction of their fellow villagers to Canadian life. The majority worked in a variety of menial, and usually demeaning, occupations. Poetry and anecdotal accounts leave us with the impression that nearly all of the Issei women worked hard, and that they found severe hardship in working. This impression is no doubt based both on the actual physical hardship endured and on the ideological contradiction between their working and the Meiji definition of the good wife and mother. Since women's work was viewed as an evil necessity, antithetical to proper raising of children, it is impossible to find accounts of work put forward with pride or a sense of accomplishment. It is common, however, to find accounts of the suffering that accompanied women's labours such as these (trans. Ito 1973, 254–81):

Tolerating dark
Days in the pit of despair ...
Well remembered past!
(Fujie)

Bank notes coming due!
Wretched hardships all endured
In those days gone by.
(Katsuko)

No respite, no rest.
Every moment filled with work.
No time for tankas,[12]
No moment all the day long.
Pitiable way to live!
(Kimiko Ono)

Both my hands grimy
Unable to wipe away
The sweat from my brow,
Using one arm as towel –
That was I ... working ... working
(Kimiko Ono)

Poems are an especially good source for understanding the experiences of the Issei, both men and women, because they were written by so many. The practice of poetry writing is a common social activity, based on the vernacular tradition of the poetry contest, in Japan. The poem is itself a social construction of reality and one of the major forms of emotional expression within the Japanese immigrant culture; it is accepted in a way that public display, or even private display in the presence of one's husband (or certainly one's children), would not be. It bears emphasizing, then, that poetry not only expresses but also structures experience (Kobayashi 1992b). It conveys common feeling and sets a standard against which the emotions of other women are, in turn, measured. The generalized experience of the Issei woman is thereby inscribed, and its validity lies in the fact that the poem was one of the important traditional vehicles through which Meiji ideology was transmitted; in the fact that the experience it depicts was actual; and in its continued importance as that which defines a woman's place.

RACIST CANADA

Although the motivations and attitudes with which young women embarked upon life in Canada can be understood almost entirely in terms of her place within the Japanese family and, at least in the earlier years, according to objectives surrounding the place of the agricultural life in the Meiji economy, their lives in Canada were severely constrained by the fact that they were considered an underclass and were generally despised by the "White" Canadian population.

The history of institutional and systemic racism through which Japanese Canadians were kept in their place in British Columbia can readily be documented (Kobayashi 1988). The first half of the twentieth century saw a series of discriminatory legislation, including the Hayashi-Lemieux Agreement, to control Japanese immigration. Japanese Canadians were excluded from the professions and were denied franchises. In Vancouver Little Tokyo, like Chinatown (Anderson 1986), was simultaneously a place controlled (and shunned) by the dominant society and a haven of institutional completeness within. Canadian-born school children were periodically denied access to public schools because of their "race" (Adachi 1968, 107; Nakayama 1983, 93–9). Occasionally violence against Japanese Canadians erupted on city streets, most notably in the riots of 1907. Newspapers during the entire period printed virulent attacks on "orientals" and warned of the "yellow peril" (Ward 1978). Wages of Japanese-Canadian workers were controlled at lower rates (British Columbia, Legislative Assembly 1927; Canada, *Sessional Papers* 1902, 1908). Finally Japanese Canadians endured one of the worst abuses of human rights in history: the forced removal, internment, and dispossession of property during the 1940s (Adachi 1976; Kobayashi 1988; Sunahara 1981). Until the 1940s, when families were split apart and the brutality of actions against them was too great to ignore, Issei women remained relatively insulated from the direct consequences of racism, not because they were not affected, but because they shut themselves away and confined themselves to their own activities, where possible, and normally allowed their husbands to represent the family publicly. A vast network of community organizations provided an institutional bulwark against the difficulties of Canadian life, and responsibility for public negotiation was clearly vested in the leaders of those organizations. Of course large areas of Canadian life – social status, education, the professions, participation in most public activities – were completely inaccessible to these women; but in order to recognize this inaccessibility, much less to challenge its authority, they would first have had to break past the patriarchal conditions that would deny such activities to them in any country or context. They were thus inhabitants of a double ghetto. Ghettoized, they were both abused and protected.

Typical of their generation, the Issei say very little in contemporary interviews, or in written accounts, of their experiences of discrimination; more often they deny the discrimination, preferring to circumscribe the boundaries of their worlds in such a way that interaction with the larger community becomes irrelevant. This denial, too, may be a factor in their tendency to deny racism in the workplace, where it most often occurred. Some of their experiences surface, however, in anecdotes about their working lives. A former waitress recalls: "when a customer left a tip once, I misunderstood that he left it by mistake, so I called him and tried to give it back. But he answered, 'It's a tip. It's yours.' In Japan I thought it was only beggars who are given small change" (Ito 1973, 285).

Whether in retrospect romanticized or, in this case, "comedized," adjustment to such incidents was one of the central features of the Issei women's life. When, in 1981, I interviewed an elderly woman in Japan who had spent some years working as a domestic servant in Canada, she rapidly became upset, expressing anger and contempt for the woman who had treated her "like an animal." Those who have remained in Canada, however, seldom express such bitterness. Instead, like the woman in the previous anecdote, they laugh at themselves, or they express gratitude for small (if patronizing) kindnesses: "I helped the farmer's wife with the housework in her home the day after arrival for twenty dollars a month which was considered 'going wages' in those days [1943]. My employer was a kind and patient woman. She taught me a great deal about housework on a prairie farm which among other things included churning butter, separating milk, baking bread, and canning jars and jars of preserves" (Moriyama 1982). It is not possible to separate racism analytically from the total flow of the lives of the Issei women, just as it is not possible to draw a distinct line between their home lives and their work lives. Racism was a fact of their existence no more deniable than was their sex; it defined the place that was Canada, and they adjusted to Canada, not in an attempt to overcome racism, but in the attempt to conform to what was expected of them as Japanese mothers and wives.

FOR THE SAKE OF THE CHILDREN

Overriding all other roles for the Japanese woman of the Meiji period was the bearing and raising of children. This was also the case for the Issei women who became Canadian, but with one major difference.

The primary justification for children in Japan had been the continuation of the heritage of the household, the honouring of the ancestors
in place, and the conformity to norms established both in tradition
and in economic exigency and founded on a universal acceptance of
patriarchy. Those women who eventually returned to Japan were able
to view their time in Canada as a sacrifice in the interests of the
household. For those who remained permanently in Canada, however,
there was no such need for continuity of the past. The birth of children
became instead a means of securing the future and, although the basic
justification may have been different from that in Japan, the social
means by which this security was achieved was remarkably similar. It
is this particular need that sets the experience of motherhood for Issei
women apart from the universal experience of motherhood. Again
poetry conveys the sense of justification for which Issei mothers
worked:

> Mothers can do all
> Required to live and endure
> In this bitter world.
> (Mitsuko; trans. Ito 1973, 279)

> For the sake of an ailing husband
> And a crippled child
> I must live.
> Take medicine for my cold
> And retire early.
> (Kimie Oda, Kisaragi Poem Study Group 1975, 149)

> By Meiji parents
> Emigrants to Canada
> The Nisei were raised to be
> Canadian citizens
> Of whom they could be proud.
> (Ko Kadonaga, Kisaragi Poem Study Group 1975, 69)

> How long I will live
> I do not know.
> I will continue to live
> Women's historical role.
> (Kinori Oka, Kisaragi Poem Study Group 1975, 156)

The sense of fate, the hint of martyrdom, and the faith in a better
future for her children are all part of the social persona of the Issei

mother. This persona is a mantle donned readily enough because it is the mantle of social acceptability. It is woven of threads spun from conditions of modernization and within the context of emigration, and it is interfaced with the rhetoric of the Meiji period, which constantly provided a source of ideological conditioning. A final poignant expression of the satisfaction it could bring is the last paragraph of Haru Moriyama's account of her life, in which the dominance of her wife/ mother role is very clear: "Weddings, babies, deaths – life goes on! Many different professions and occupations are represented in our big family. I am proud that I did my share in giving Canada some good citizens. Although the family has gone through many changes ... the thread of goodness that has prevailed in the Moriyama family remains the same. Toyoki [her husband] always maintained that hard work never killed anyone. He believed in honesty and self respect. He believed in helping his fellow man. These values have rubbed off on every member of my family. For this, I am truly grateful" (Moriyama 1982).

CONCLUSION

Late nineteenth-century Japanese farmers, men and women, repatterned their lives to accommodate a system of political economy that set new standards of social value and legitimation in the interests of national development. When these standards were undermined by economic exigency, the farmers turned to traditional structures of patriarchy to effect an acceptable adaptation to both the dominant political ideology and the economic means of achieving their goals. The industrial economy and the patriarchal society thus reinforced each other, as the interests of modernization and national development became the interests of the household. But the tradition was a contemporary creation that drew upon past structures to accommodate the present, especially with regards to the practices of marriage and the definition of relations within the household. Tradition was remade most quickly at the interface of home and work, where new patterns of labour became a means of preserving agrarian life. Emigration to Canada was undertaken within this context as a means of supporting and further strengthening that system, of achieving what was most valued but least accessible under Meiji conditions. In Canada, partly in spite of and partly in defense against a context of racism, the dominant Japanese ideology continued and provided for many a paradoxical means of reentering the now deeply conservative agrarian village through industrial waged labour. Their marginalization in Canadian society served, if anything, to reinforce traditional values, including patriarchy, as those things that were attainable in an old place and time but impossible in the new place and time of Canada. Even for those who chose not to return to Japan, the ideal definition of the household and the structure of the family changed little until the second generation.

Haru Moriyama's words were written in 1982; she died a year later. But her expressions of Meiji values – a loyal wife and good mother, who puts the welfare of her children and the honour of her husband first – echo those of nearly a hundred years before. Her sense of tradition was a construction in the present that justified her life retrospectively. This example illustrates the double irony of Issei women's lives. Coming to terms with life and work in Canada was secondary to coming to terms with what was expected from a woman of Japan. Therefore, despite the variety of circumstances in which Issei women found themselves and despite the variety of their responses to those circumstances, there was a common structuring of the ways in which they responded to the rapid and wrenching changes in their lives; thus they accommodated, if not overcame, the intractible contradictions set by the life into which they had been plunged. That commonality is linked to the patriarchal structure of Japanese life, which functions as both the means of oppression and the source of socially defined happiness. A woman, despite her fears, submitted to the demands of husband, father, or in-laws, including the demand to work outside the home in order to support the household. But it was the role of mother, as reproducer in and of the patriarchal system, that gave her the greatest gratification. And ultimately the difficulties of being cast into the Canadian labour force were cast aside by the satisfaction of motherhood.

The pattern of Issei women's experience provides an opportunity for analysis according to three objectives on the feminist agenda for understanding women's work. The first is to understand the relation between home and work as complex and interconnected, so that it is not only impossible to define them as separate places but that, within a broader context of the political economy, change in one is seen to instigate potential change in the other. The "rules" that structure this relationship are thoroughly embedded in cultural practices, of course. The second objective is to understand the nature of patriarchy not only as the systematic subordination of women, powered by official means as well as by the demands of community pressure, but also as a construction that is highly charged with emotion, that is at the basis of the most intimate relationships that define a woman's place, and that is the source of her happiness. To recognize patriarchy as a source of happiness for the Issei is not to recommend it as a form of social

relations, for, as we have seen, the contradictions engendered in patriarchal relations often cause anything but happiness for the women
who face them; rather it is to recognize that part of the strength of
patriarchy lies in its built-in capacity to create happiness through
subordination to a hegemonic ideal – to provide, in Connell's (1987,
187) terms, the "option of compliance." The third feminist goal is to
recognize that the social construction of tradition is a project of the
present that remakes the past on its own terms and that provides a
stable vehicle for change in specific times and places. If the legitimacy
of the past is often illusory, its expression is nonetheless grounded in
the activities and the setting of daily life in the present.

NOTES

This work was supported by grant 410-86-0644 from the Social Sciences
and Humanities Research Council of Canada.

1 Several works analyzing the Meiji economic transition are now viewed
   as classics: Allen (1972); Nakamura (1966); Norman (1940); Smith
   (1955); Lockwood (1965, 1968). For a discussion of agrarian life in the
   Meiji period, see Irokawa (1985); for the effects of conscription, Crowley (1966) and Norman (1943); for conditions among new factory workers, Allinson (1975) and Wilkinson (1965); for population change,
   Tauber (1958); for entrepreneurship, Hirschmeier (1964); for education,
   Dore (1964). For discussions of social conflict that arose as a result of
   Meiji policies, see Bowen (1980); Hane (1982); Najita and Koschmann
   (1982). The most comprehensive work on Meiji ideology is Gluck
   (1985). For a review of case studies of the effects of Meiji policies on
   farmers' lives, see Kobayashi (1983). For a general history of the
   period, Hane (1986) is recommended.

2 The reverse was also true for the many men who married into families
   that had no sons (yoshi). The difference, of course, is that men thus
   became household heads in their own right, while women maintained a
   subordinate position, no matter what their origin.

3 Patricia Tsurumi's (1990) Factory Girls provides a graphic and poignant
   description of conditions in the thread mills.

4 There is only one other case study conducted at the village scale that
   can be used for comparison. Ishikawa (1967) also finds that the first
   emigrants were heads of households and first sons, followed by second
   and third sons. In this study, however, the rate of successful reestablish-
   ment in the village was much lower. The difference may lie in that Ishikawa studied emigrants to Hawaii who worked for poor wages in the
   plantations, whereas the Kobayashi study considered workers who fared

relatively much better in Canada. A comprehensive comparison of regional emigration trends is currently being conducted by the author.

5 Ironically this attitude backfired against Japanese Canadians in the 1940s, by which time the majority of the community was Canadian-born. Racist actions against them, including uprooting, internment, and dispossession of property and human rights, were justified on the grounds that they must be prevented from further breeding (Kobayashi 1988). Chinese Canadians, on the other hand, were treated much differently. Immigration of Japanese men was curbed by a prohibitive head tax of $500, and women were forbidden to enter the country at all (Wickberg et al. 1982).

6 Interviews were conducted with returned emigrants in 1981–82 and again in 1986. Detailed results of the interviews are not provided here.

7 For personal accounts of the experiences of picture brides, see Ito (1973, 188–201); the Japanese-Canadian Centennial Project (1978, 18–19); Moriyama (1982); Nakayama (1982, 108–10).

8 Source: *Gaimusho Gaikoshiryo Toshokan* [Department of External Affairs Foreign Records Library], *Kaigai Zairyu Dohojin Shokugyobetsu Jinko Chosa* [*Survey of the population and occupations of Japanese abroad*], K.3.7.0.7 to K.3.8.2.5–7.

9 Nolte (1983) interprets this law as an indication that it was social circumstances, not biology, that formed the logic of Japanese patriarchy.

10 In contrast to the Canadian data, consular reports from Seattle show somewhat larger proportions of women working in domestic service. Nakano-Glen (1986), in a book on this subject, relies on census data to find that just over half of all Issei women worked as domestic servants. Whereas it is possible that the figure is legitimately lower in Canada, especially since so many Canadian women worked in the fish canneries, it seems implausible that it was as low as the consular reports indicate.

11 This is part of a two-volume set (Nakayama 1921a, b) containing 3,275 pages that describe every aspect of Japanese-Canadian life and provide information about Canada that would be of use to new immigrants. Information for the book was collected over the period 1917–19 by Jinshiro Nakayama, a reporter for a Tokyo newspaper, who travelled throughout British Columbia visiting Japanese immigrants. The biographies are one-page entries, usually quite effusive in their praise of the "successes" of the immigrants. Interviews today indicate that the author charged a listing fee to the 700-odd individuals, and so the sample is biased in favour of those with the money and the desire to see their names in print. Nonetheless, there is no richer source of information on the conditions of life for early Japanese immigrants.

12 A tanka is a five-line Japanese poem, as presented here.

REFERENCES

Adachi, Ken. 1976. *The Enemy that Never Was: A History of Japanese Canadians*. Toronto: McClelland and Stewart.

Alexander, Sally and Barbara Taylor. 1981. "In defence of 'patriarchy'." In *People's History and Socialist Theory*, R. Samuel, ed, 370–3. London: Routledge and Kegan Paul.

Allen, George Cyril. 1972. *A Short Economic History of Modern Japan, 1867–1937 with a Supplementary Chapter on Economic Recovery and Expansion, 1945–1970*. London: Allen and Unwin.

Allinson, Gary. 1975. *Japanese Urbanism: Industry and Politics in Kariya, 1872–1972*. Berkeley, Los Angeles and London: University of California Press.

Anderson, Kay. 1986. "East" as "West": place, state and the institutionalization of myth in Vancouver's Chinatown, 1880–1980. PhD diss., University of British Columbia.

Bacon, Alice. 1902. Rev. ed. *Japanese Girls and Women*. Boston and New York: Houghton Mifflin. Orig. pub., 1891.

Befu, Harumi. 1971. *Japan: An Anthopological Introduction*. Tokyo: Tuttle.

Bernstein, Gail. 1976. "Women in rural Japan." In *Women in Changing Japan*, J. Lebra, J. Paulson, and E. Powers, eds, 25–50. Stanford: Stanford University Press.

Bowen, Roger. 1980. *Rebellion and Democracy in Meiji Japan: A Study of Commoners in the Popular Rights Movement*. San Francisco, Los Angeles, London: University of California Press.

British Columbia. Legislative Assembly. 1927. *Report on Oriental activities within the province*. Victoria.

Canada. *Sessional Papers*. 1902. *Report of the royal commission by which Oriental labourers have been induced to come to Canada*. Ottawa.

– 1908. *Report of the royal commission on Chinese and Japanese immigrants*. Ottawa.

Crowley, James. 1966. "From closed door to empire." In *Modern Japanese Leadership: Transition and Change*, Bernard S. Silberman and H.D. Harootunian, eds, 261–88. Tucson: University of Arizona Press.

Dore, Ronald. 1964. "Education: Japan." In *Political Modernization in Japan and Turkey*, R.E. Ward and D.A. Rustow, eds, 176–204. Princeton: Princeton University Press.

Gluck, Carol. 1985. *Japan's Modern Myths: Ideology in the Late Meiji Period*. Princeton: Princeton University Press.

Hane, Mikiso. 1982. *Peasants, Rebels and Outcasts*. New York: Pantheon Books.

– 1986. *Modern Japan: A Historical Survey*. Boulder and London: Westview Press.

Hartmann, Heidi I. 1979. "The unhappy marriage of Marxism and feminism: towards a more progressive union." *Capital and Class* 8: 1–33.

Hendry, Joy. 1981. *Marriage in Changing Japan.* Tokyo: Tuttle.

Hirschmeier, Johannes. 1964. *Origins of Entrepreneurship in Meiji Japan.* Cambridge, Mass.: Harvard University Press.

Irokawa, Daikichi. 1985. *The Culture of the Meiji Period.* M.B. Jansen, trans. Princeton: Princeton University Press.

Ishikawa, Tomonori. 1967. "*Yamaguchiken Ooshimagun Kugason shoki keiyaku imin no shakai chirigakuteki koosatsu*" [A social geographical study of contract labour from Kuga, Oshima, Yamaguchi]. *Chirigaku* 7: 25–38.

Ito, Kazuo. 1973. *Issei: A History of Japanese Immigrants in North American.* S. Nakamura and J.S. Gerard, trans. Seattle: Executive Committee for the Publication of *Issei.*

Japan. Dept of External Affairs. 1893, 1910, 1920. *Annual consular reports.* Tokyo.

– 1992. *Survey of the population and occupations of Japanese abroad.* 17 vols. Foreign Records Library.

Japanese-Canadian Centennial Project. 1978. *A Dream of Riches: The Japanese Canadians, 1877–1977.* Vancouver: The Japanese Canadian Centennial Project.

Kawashima, Takeyoshi and Kurt Steiner. 1960. "Modernization and divorce rate trends in Japan." *Economic Development and Cultural Change* 9. 1: 213–40.

Kisaragi Poem Study Group. 1975. *Maple Tanka Poem by Japanese Canadians.* Toronto: The Continental Times.

Kobayashi, Audrey. 1980. "Landscape and the poetic act: the role of haiku clubs for the Issei." *Landscape* 24. 1: 42–7.

– 1983. Emigration from Kaideima, Japan, 1885–1950: an analysis of community and landscape change. PhD diss., University of California at Los Angeles.

– 1984. "Emigration to Canada and the development of the residential landscape in a Japanese village: the paradox of the sojourner." *Canadian Ethnic Studies* 16. 3: 111–31.

– 1985. "Emigration to Canada, landholding and social networks in a Japanese village, 1885–1950." in *Japanese Studies in Canada,* M. Soga and B. St. Jacques, eds, 162–86. Ottawa: Canadian Asian Studies Association.

– 1986. Regional backgrounds of Japanese emigrants to Canada and the social consequences of regional diversity for Japanese Canadians. Albatross Discussion Paper, no. 1. Montreal: McGill University, Department of Geography.

– 1988. The law as justification of spatial tyranny: the case of Japanese Canadians. Paper presented at the annual meeting of the Canadian Law and Society Association, 7–9 June, Windsor, Ontario.

– 1992a. *Memories of Our Past: A Brief History and Walking Tour of Powell Street*. Vancouver: NRC Publishing.

– 1992b. "Structured feeling: Japanese-Canadian poetry and landscape." In *A Few Acres of Snow: Literary and Artistic Images of Canada*, Paul Simpson-Housley and Glen Norcliffe, eds, 243–57. Toronto: Dundern Press.

Koyama, Takashi. 1962. "Changing family structure in Japan." In *Japanese Culture: Its Development and Characteristics*, R.J. Smith and R.K. Beardsley, eds, 7–54. Chicago: Aldine.

Lebra, Takie Sugiyama. 1984. *Japanese Women: Constraint and Fulfilment*. Honolulu: University of Hawaii Press.

Lockwood, W.W., ed. 1965. *The State and Economic Enterprise in Japan: Essays in the Political Economy of Growth*. Princeton: Princeton University Press.

Lockwood, W.W. 1968. *The Economic Development of Japan: Growth and Structural Change*. Princeton: Princeton University Press.

Moriyama, Haru. 1982. Rambling reminiscences of Haru Moriyama, recorded and expanded by her daughter, Fumi Tamagi. Unpublished document.

Najita, Tetsuo and J. Victor Koschmann, eds. 1982. *Conflict in Modern Japanese History: The Neglected Tradition*. Princeton: Princeton University Press.

Nakamura, James. 1966. *Agricultural Production and the Economic Development of Japan, 1873–1922*. Princeton: Princeton University Press.

Nakamo Glenn, Evelyn. 1986. *Issei, Nisei, War Bride: Three Generations of Japanese American Women in Domestic Service*. Philadelphia: Temple University Press.

Nakayama, Gordon. 1982. *Issei*. Toronto: Brittania.

Nakayama, Jinshiro. 1921a. *Kanada no Hoko* [The treasure of Canada]. Tokyo: Nakayama.

– 1921b. *Kanada Doho Hatten Taidan* [The Encyclopedia of Japanese in Canada]. Tokyo: Nakayama.

Nolte, Sharon H. 1983. Women, the state and repression in Imperial Japan. Working paper, no. 3.

Norman, E. Herbert. 1940. *Japan's Emergence as a Modern State: Political and Economic Problems of the Meiji Period*. New York: International Secretariat of the Institute of Pacific Relations.

– 1943. *Soldier and Peasant in Japan: The origins of Conscription*. New York: International Secretariat of the Institute of Pacific Relations.

Paulson, Joy. 1976. "Evolution of the feminine ideal." In *Women in Changing Japan*, J. Lebra, J. Paulson, and E. Powers, eds, 1–24. Stanford: Stanford University Press.

Robins-Mowry, Dorothy. 1983. *The Hidden Sun: Women of Modern Japan*. Boulder: Westview Press.

Rowbotham, Sheila. 1981. "The trouble with patriarchy." In *People's History and Socialist Theory*, R. Samuel, ed, 364–9. London: Routledge and Kegan Paul.

Shibata, Yuko, Shoji Matsumoto, Rintaro Hayashi and Shotaro Iida. 1977. *The Role of Japanese Canadians in the Early Fishing Industry in BC and an Annotated Bibliography. The Forgotten History of the Japanese Canadians, volume 1.* Vancouver: New Sun.

Smith, Thomas. 1955. *Political Change and Industrial Development in Japan: Government Enterprise, 1865–1880.* Stanford: Stanford University Press.

– 1970. *Agrarian Origins of Modern Japan.* Stanford: Stanford University Press.

Steveston Imperial Cannery Collection. 1913. Vancouver Public Library, Vancouver, BC.

Sunahara, Ann. 1981. *The Politics of Racism: The Uprooting of Japanese Canadians During the Second World War.* Toronto: Lorimer.

Tauber, Irene. 1958. *The Population of Japan.* Princeton: Princeton University Press.

Tsurumi, E. Patricia. 1990. *Factory Girls: Women in the Thread Mills of Meiji Japan.* Princeton, NJ: Princeton University Press.

Ward, Peter. 1978. *White Canada Forever: Popular Attitudes and Public Policy Toward Orientals in British Columbia.* Montreal: McGill-Queen's University Press.

Wickberg, Edgar et al. 1982. *From China to Canada: A History of the Chinese Communities in Canada.* Toronto: McClelland and Stewart.

Wilkinson, Thomas O. 1965. *The Urbanization of Japanese Labour, 1868–1955.* Amherst, Mass.: University of Massachusetts Press.

4 Womanly Militance, Neighbourly Wrath: New Scripts for Old Roles in a Small-Town Textile Strike

JOY PARR

In the years immediately following World War II several Canadian unions undertook major organizing campaigns among women workers in secondary manufacturing and the retail trades. These initiatives were largely unsuccessful, and thereafter labour organizations returned their attention to male workers. Only relatively recently has the imperative to recruit members in female-dominated sectors reemerged in the labour movement. This study is part of a larger historical project that compares a community of unionized male furniture workers with a community of female hosiery workers. A central tenet of this research is that workplace relations of both female and male workers were more profoundly formed in terms of gender than heretofore has been acknowledged. For the lone strike in the hosiery town, the discussion will explore the ways in which militance was gendered. Apparently women formulated their grievances, claimed the right to act, and acted differently from men during this dispute; yet this distinction was not well recognized at the time and has become more elusive still since the community, with passing years, has reshaped its collective memory of this traumatic event. Yet the differences between women and men in the origins, instruments, and perceptions of militance bear close attention. In apprehensions and misapprehensions of womanly militance, both contemporary and retrospective, the force of female unionists' activism in the hosiery strike was muted, deflected, and lost. By examining why womanly militance was so discomfiting for male observers to acknowledge and for female participants to claim, we help reopen the possibility of vital and effective union organizing among women workers.

On 18 January 1949 the members of Local 153 of the United Textile Workers of America (American Federation of Labor [AFL]), which had been certified the previous autumn as the bargaining agent for the employees of Penman's Company in Paris, Ontario (population 5,000), went out on strike against the firm. So searing was the experience that thirty-five years later townspeople remembered with unfailing accuracy who had struck and who had crossed the line; house by house, block by block, neighbours were known according to which side they had been on during during those three months.[1] Until the last of the knitting mills in town closed in the autumn of 1984, the women and men who worked for Penman's foreswore that the strike was a tactic and kept a wary distance from labour organization generally.

It was not the formal failure to gain contractual demands that gave the strike its power in memory. The pained and unforgiving recollections of those weeks are rather of individual lines spoken and postures struck, of defiant gestures and submissive acts; it is the theatre of the strike that is burned in memory – the choreography in public view, the solitary soliloquy delivered in the back kitchen late at night. The strike put accustomed social relations in town fundamentally in question and cast townspeople in roles they had never rehearsed and never intended to play. The words mother, sister, friend, neighbour, boss, foreman, constable, and court were emptied of their conventional meanings and found to entail obligations, demand behaviours, and convey values not previously acknowledged or understood. The strike is remembered as a time when the whole town was transformed into an alien territory, where all events were like those on a stage: grotesque, harshly shadowed, and played toward extremes. Those who remember are accepting of a dissonant undercurrent in their narrative; they tell their stories as if they had lived their lives out of bounds in those months, yet as if an implacable force had informed and ordained their every act. They pause longest to consider the roles of woman and neighbour, because the strike magnified difference in identities constructed by gender and community and played upon the conflicting possibilities of release from and restraint of those identities.

In many settings womanly militance might seem an oxymoron, and labour activism among small-town women, an impossibility. But the history and traditions of Paris, Ontario, were rich in the resources from which effective collective action might be crafted. The assisted emigration schemes that Penman's sponsored until 1928 recruited women for their skills as hosiery workers and for the traditions in their home districts of life-long female waged work; the emigrants joined a community where millwork was much more likely than marriage to provide the sustaining continuity in an adult woman's life. For half a century and more the female millworkers of Paris had managed waged work and motherhood by relying upon elaborate kin and neighbourhood networks of support, which made moot the boundaries between household and community, market and nonmarket labour and extended the collectivity of interested parties in any workplace dispute far beyond the limited numbers on the biweekly Penman's payroll. The paternalistic practices for which the firm was renowned – the company-sponsored YWCA, the company-owned housing, the recreational association, the annual lakeside outings, and the discretionary pension plan – expanded the common interests Penman's employees shared. But these collective sensibilities were the stuff from which both dogged loyalty and determined opposition might be made (Parr 1987b, 529–51; Parr 1987a, 137–62).

Paris was unconventional in its demography, labour force, family structures, and gender roles; its honoured and reflexive local habits of mind distinguished the community from the mainstream culture. As Harold Benenson (1985, 121–8) and Jacquelyn Dowd Hall (1986, 356–7) have emphasized, industrial composition and family and community structures mediate workers' protests in cases such as this. In communities of women workers, cross-class and cross-generational alliances give women's activism force, while regional understandings of sexuality and female friendship give womanly militance meaning. We have gone beyond conventional wisdom concerning female distractedness or acquiescence in the workplace to a more complex, less essentialist understanding of the relationship between gender and militance. At issue now is not the existence of female activism but "the conditions under which women's participation in collective action will be more or less successful" (Waldinger 1985, 87–8), conditions that give rise to distinctive organizing opportunities and inhibitions within communities (Santos 1985, 230–1; Turbin 1987, 47–50).

Moments of labour crisis, however, also prompt invasions: of senior executives, labour organizers, representatives of national manufacturers associations, and trade union federations, who come to manage the dispute; of state apparatus crafted beyond the local level – police, judges, and lawyers, who arrive to contain and ajudicate; of journalists and photographers, who descend in the moments of most acute conflict to interpret and display the event for their distant audiences. Within the community the blinding stage lights and the commanding, externalizing attention are disorienting and intimidating, and prone to call into question local confidence in local ways. In the case of Paris the dispute became an unequal dialogue between the community and the mainstream culture in ways that were particularly corrosive to the distinctive gender identities the town had both accommodated and nourished.

PROLOGUE

The UTWA had begun organizing in Ontario in the spring of 1944. By 1946 it had established eleven locals in the province; seven were in eastern Ontario, and five of these within easy reach of Paris in the communities of Brantford and Woodstock. At the end of the war earnings of hosiery and knit goods workers were the lowest in the textile sector. The average manufacturing employee took home 50 per cent more than the average hosiery worker. The UTWA achieved their greatest gains, both in wages and union security, in woven goods plants, where prices and profits were rising more rapidly than in knit goods. But organizers in the hosiery sector could take heart from their successes at the York Knitting Mills in Woodstock, Ontario, 30 kilometres west of Paris, where, by November 1946, Local 125 had achieved a considerable measure of union security (through a maintenance of membership clause), as well as two weeks' paid vacation and a regular schedule of wage increases (Parent-Rowley Collection, 1/15, 5/5).[2]

In Paris the UTWA faced an uphill battle from the start. Jerry Regan, a veteran organizer among male furniture workers, sized up the community after his first weeks there in the spring of 1946 and described it as "one of the worst of the company towns." There was a core of support for the union, which coalesced quickly. By late April one quarter of the employees in the hosiery and underwear mills and two-thirds of those in the sweater mill had signed cards. The lowest response was in the yarn mill, where an earlier attempt to organize a spinners' union had been thwarted (Interview, Frances Randall) and where a works council was in place. The local was chartered in June but by July organizers were worried: the progress of the membership drive had been halted; the members of the works council had been reconstituted as a company union; and a vigorous campaign had begun to discredit the American Federation of Labor as a "foreign labour invasion" and the UTWA as communist led. "The problem has become political rather than union," Regan reported; "the company have decided they can beat us." In an attempt to consolidate its threatened position, the local applied for certification before the Ontario Labour Relations Board in August, but the application failed.[3]

Throughout the spring of 1947 the twelve members of the local's executive, led by Charles Alexander, a boarder from the hosiery mill, tried to hold their ground. They were "hoping for something to give us a break," and were discouraged by how "indifferent and self-satisfied" their coworkers seemed to be. During the year the campaign was conducted under a dark pall (Parent-Rowley Collection, 2/10).[4]

The UTWA had been organizing and leading strikes among textile workers in Quebec, where Penman's and Dominion Textiles, a woven goods firm with which Penman's had close corporate links, had several large plants; and Kent Rowley and Madeleine Parent, members of the Canadian executive of the union, had been charged with seditious conspiracy in connection with a Quebec strike. The revelations concerning espionage of Igor Gouzenko, a Russian embassy cipher clerk who had defected to Ottawa, heightened fears of communist infiltration in all areas of Canadian life. The widely reported trials of Rowley and Parent focused centrally on allegations that each was, or had been, a member of the Communist Party. In Paris both the Canadian Textile Workers, which was the company union, and the CIO-affiliated Textile Workers of America, which was competing with the UTWA to organize plants in southern Ontario, adopted anticommunist rhetoric. Several of the UTWA organizers were party members, most notable among them William E. Stewart, who led the strike in Paris in 1949 and later became leader of the Ontario Communist Party. Anthony Valente, American president of the UTWA, long suspected the politics of his Canadian staff and in 1952 would fire twelve of them, including Rowley and Parent, on these grounds.

Such a concerted and sustained attempt to organize against an employer was unprecedented, even revolutionary, in Paris (Smith 1980). Many ordinary millworkers separated the workplace goals of the UTWA from whatever wider political agenda the union was said to espouse. Several prominent community leaders, among them Martin Hogan and Lawrence Brockbank, from the school board, and Donald Smith, a teacher and later principal at the high school, called the communist label "a good tactic" used to prejudice legitimate demands at the mill (Historical Perspectives Collection).[5] It was indeed. Because union membership was uncustomary in town, its local benefits still prospective and unproven, and because the notions of demand and bargain were so different from the established patterns of request and wait, the revolutionary teleology that was drawn out by rival unions and the firm made an event that had already caused considerable uncertainty seem even more formidable.

In the spring of 1948, Local 153 applied successfully for certification.[6] A three-man conciliation board was established in September and reported on 12 November. The majority conciliation report recommended a 5 cent increase and the very slim union security of the "voluntary revocable check-off," agreeing with company claims that the leadership of the UTWA was "irresponsible" and had shown itself entitled to no greater form of protection. The minority report was written by Drummond Wren of the Workers' Education Association, who was the union's nominee to the board; he recommended a 15 cent hourly increase and, noting the continuing close associations between the company and the employees' association, a maintenance-of-membership clause along with the provision that all new employees be required to become members of the union. The company granted the 5 cent increase in late November, and thereafter declined to negotiate.[7] In a letter of 27 October to Drummond Wren, William Stewart, the field representative of the UTWA in Paris, used the diction of the tank commander he had lately been: "We are making preparations to pull the pin in the event that this conciliation does not effect a settlement" (Parent-Rowley Collection, 11/6).[8] Within the community general opinion on the best next step was less decided.

THE CAST OF UNION-MINDED

Of the 693 workers on Penman's payrolls in 1948, 56 per cent were women; 42 per cent of the 433 members of Local 153 were female (Personnel Records).[9] The union's own organizing structure may in part account for the lower representation of women among union members. The first female UTWA organizer in town, Helen McMaster Muller, arrived only after the strike vote had been held; Penman's workers were recruited through evening home visits, and male field representatives of the union may have been less than vigorous or even at ease in their approaches to the many all-female households in the community.[10] Still, female militance in town was considerable. When the strike call finally came, 44 per cent of the those who responded by not crossing the line were women.

There were stark differences in the personal characteristics of female and male strikers. Women who supported the strike were on average six years older than the men and four years older than average among women employed by the firm. Single men were overrepresented among the strikers; among women the pattern was reversed. There were markedly more wives, and especially widows, among the strikers than in the work force as a whole.[11] It is not surprising that older women, wives, and widows were most conspicuous among the female union activists. In Paris, as in the textile centres of New York and New England and the woven goods districts of northern France, these women had the most compelling and long-standing committment to waged work (Tilly 1981, 415–17; Cameron 1985, 42–61). "Over the years they'd been kinda ripped off," a senior male worker recalled (Interview, Sam Howell); "this was a chance for them." The oldest striker, reportedly, was sixty-two-year-old Florence Miller, a winder who lived with her eighty-two-year-old husband in a rented house in town (*Brantford Expositor*, 28 February 1949). While in their young years female workers had been fearful of the boss, fear turned to disdain as they took responsibility for raising children and running households. They grew less tolerant of the petty tyranny and favouritism practised by the foremen and lead hands, more confident of their own worth, more confirmed in their own sense of dignity and honour (Kessler-Harris 1985, 119; Frankel 1984, 52; Interviews, Charles Harrison, Robert Fletcher; Smith 1980, 20–1, 40). "I didn't want to feel helpless; I wanted my rights," Lottie Keen (Interview), a skilled looper and a widow raising her son alone, remembered; "I thought it would give more security to people. That is what I understood – unions were for the workers." Betty Shaw's mother (Interview), an English woman who brought her kin to join her in Paris and who by 1949 had worked thirty years in the mill, thought the strike was about fairness and a way to protect her daughter from workplace conventions that she despised.

The sense among Paris workers of generations passing manifested itself as a perceived investment in and claim on the firm. "I remembered how tired my mother was back about 1913 after she worked all day and then had to come home and get dinner … They had always taken too much out of us … I got up at that meeting," Florence Lewis said quietly (Interview), her voice shaking slightly at the recollection, "and I said I favoured the strike to make things better for my children. You wouldn't believe what it was like when I was a teenager in 1925. Everybody wants things better for their children." Long years working side-by-side in the mill with neighbours and kin forged a bond of common predicament. Betty Shaw (Interview) said of her mother and her elderly aunt, both of whom were fired after the strike, "Everybody thought they had better stick together … I don't believe they ever would have retired as long as they were able to go." For Mildred Hopper (Interview), a seamer in the sweater mill, the dispute was not about money: "Myself, I figured I wasn't making a bad wage, but then I thought well if you don't stick together you are not going to get any place either. I think it was mostly to have somebody to go to and have some systems. I think it was just the principal of the thing of all being together to be able to do something." In earlier years this sense of needing "to stick together" had sustained informal practices to "help one another out" on the shop floor and extensive networks of exchange among households within the community. Hopper herself had worked away from Paris in unionized heavy industry during the war; by the late forties other female millworkers had brothers and husbands who were employed in union shops in nearby centres. To many, joining the union seemed a good way to extend and formalize long-standing habits of solidarity in town (Interviews, Lottie Keen, Mildred Hopper, Betty Shaw, Florence Lewis, Clarence Cobbett, Thomas Blaney, Robert Fletcher).

COMMUNITY: ACCESS TO THE STAGE

The feeling of belonging, of having cast in one's lot with the town; living within community traditions, by the scrupulous reckoning of obligations that were honoured and betrayed; living out community sanctions – all of these conditions of life in Paris gave townspeople cause to become engaged in the Penman's dispute, either for or against the strike. Those same habits of solidarity that made older women in town "union-minded" formed a clear boundary between insiders and outsiders in the community. They discredited the advice of strangers and gave pause to neighbours about to break rank; they also gave force to criticisms of the masters of the mill and courage to plain folk who had chosen their ground.

The UTWA was not an institution indigenous to the town; its organizers had not come to live in the community. While the rival unions railed against the AFL affiliate as an "emissary from Moscow" and "an American invasion," it was the trespass across a boundary nearer to home that caused most disquiet within town. Even among those committed to the UTWA because "an outside union could better bargain for our working condition," there was an undercurrent of unease, a worry that the organizers were "sharp-talking fellas" (Parent-Rowley Collection, 2/10) who did not listen well enough, who did not value, and thus could not be trusted to preserve, important community traditions that the work action put at risk.[12] When, at the invitation of union organizers, men from unionized metalworking plants in nearby Brantford arrived at Paris rallies to raise morale and boost confidence, they instead heightened the apprehension of danger. Their cheers of support were heard as "hollering" and "hatred" (Interviews, Ida Pelton, Paul Nelles, Irene Cobbett, Lottie Keen, Betty Shaw; see also Santos 1985, 247–9). "In a small town like Paris everybody knows everybody else you see, especially if you have been working together" (Interview, Alice Smith). Charles Harrison (Interview), weighing the implicatons of a strike upon neighbours, resigned from the union after the strike vote was taken: "That small town bit of knowing what everybody did, and who everybody was, had repercussions. It became more intense when you had a difference of opinion. It was so intermixed with the life that you were living that it would never be the same again. It's because you were so close that once you split you would split wide open." Yet loyalty to the community also steeled the resolve of women and men committed to the strike.

Tommy Curry, a long-time town resident but not a Penman's employee, claimed that unionists were motivated by "a lively sense of responsibility towards the community welfare," and that they wanted Paris to be "more than a barrack-room for industry" (*Brantford Expositor*, 22 February 1949). When Bruce Wilson, a carder from nearby Princeton with only twelve months' experience in the mill, challenged Curry's right to speak about the strike (*Brantford Expositor*, 24 February 1949), Grace Hockin replied: "My family has been in and out of Penman's for the last 30 years. I myself started in the mill in 1938 for 18 cents an hour and in the past 10 years have put in seven years in the mill ... Does Mr Wilson realize that he or maybe his co-workers are only seasonal workers, whereas the people in town have to rely on the Penman mill 365 days a year?" (*Brantford Expositor*, 1 March 1949). As a nonstriker told local historian Donald Smith, "A lot of us owned our own homes. The company encouraged us to buy ... Many felt that since they couldn't leave Penman's the best thing they could do was join the union and try to get higher wages" (Smith 1980, 22). "Remember," said Florence Lewis (Interview), a home-worker who ran a soup kitchen for the pickets from her back porch, "we were all citizens of Paris."

The union organizers were not the only outsiders in town. By the late forties the older Penman's executives in Paris, men active in municipal politics and community organizations and frequently seen about the streets and the mill, were replaced by personnel transferred from plants in Quebec who were not "town men." The feeling was widespread, especially among veterans, that the management had grown calloused, divorced from community concerns and preoccupied with profit, and that the new men were not following the principles of fair play for which they had fought overseas (Interview, Charles Harrison; Historical Perspectives Collection, interviews, Martin Hogan and Lawrence Brockband).[13] For many in town the strike was a way to sustain the structural integrity of the set, to insist that the community should be more than "a barrack-room for industry."

But there were no local instruments with which to secure this claim. Both the cause of and the apparent remedy for the town's problem lay outside community convention. The union and the firm each were making decisions in Paris on the basis of industrywide considerations. The more experience Penman's management had with the UTWA in Quebec, the more determined it was not to allow the union to gain a foothold in Paris; and the more successful UTWA organizers were in gaining union security and pay raises in other Ontario plants, the more convinced they were that Penman's, at least economically, could afford to yield to their Paris demands. In the final days, for both union and management, the issue came down to union security, a concept of industrial legality that was little understood in town. Without some effective measure of union security the company could continue to count on community loyalties to erode UTWA support. But a strike to gain union security entailed a greater risk for the community than for the union, on a issue most meaningful to the outside organizers on the basis of their experience in centres far away from Paris.[14] Charles Harrison, a skilled worker and veteran with strong views against Penman's management, became disaffected when, after the conciliation report, the union organizers "just told us that this was the way it was going to be, as though we didn't know what we should be doing." He and ten other senior workers resigned from the union (Interviews, Charles Harrison, Frances Randall, Horace Timpson, Thomas Blaney). Martin Hogan, a thoughtful community leader with a sceptical view of Penman's tactics, judged the strike ill-considered: "an interested observer could see that they were making a terrible mistake by not accepting what they could get and then organizing from inside" (Historical Perspectives Collection, interview, Martin Hogan). Like many Paris men, speaking of that time, Charles Harrison (Interview) favoured military metaphors: "The bullets were made by the people in Paris, but they wanted to hold the line before the bullet was fired, or they wanted to be the ones to fire the bullet themselves when the time came, but somebody beat them to the gun and pulled the trigger, and they didn't like that – some of them didn't like it." Seven weeks after the strike vote was taken, the strike began.

READINGS OF WOMANLINESS AND MILITANCE

The boundary around the community was plain and public; to each opinion or piece of advice offered concerning the dispute, the label "insider" or "outsider" was readily affixed. Acts and spoken thoughts were assigned citizenship, and authority was drawn from the place of residence of the person who displayed them. Citizenship might have been contested in private, but this dispute, like the boundary, was arrayed in plain and public view.

The attribution of gender to acts and speech was less straightforward. Gender was a label; it was also a screen. Thus it not only sorted behaviour as manly or womanly but also obscured or removed certain ways of being from view. Gender influenced not only what could or would be done but what could be seen and was said to have been done. It was both a mirror and a mask. The gender roles enacted for the duration of the strike were simultaneously assumed and ascribed, and the actress often found she had little control over how her part was perceived.

Respectability was the touchstone of womanly authority of the era. As Ellen Ross (1985, 39) has argued, women "*embodied* respectability or the lack of it, in their dress, public conduct, language, housekeeping, childrearing methods, spending habits, and, of course, sexual behaviour." Respectability might once have been based on gentility, part of the regalia of a particular class, but it had become equated with womanliness and woven into the whole cloth of the gender. Respectability was a virtue that resided in the self-image of women as mothers and homemakers, but that also formed their sense of their rights and responsibilities beyond the domestic sphere. It was a measure of conduct marked most in child rearing (Kessler-Harris 1985, 3–9; Jameson 1977, 171–2; Jones 1984, 449). Consider this response by a working-class woman in Paris to the "sharp-tongue" of a merchant's wife: "I terrifically resent the way she snubbed me. Our family has always been respectable. That snob may have more money than us, but we're every bit as good. At least we have smarter and better-looking kids. They're better behaved too" (Smith 1980, 65). Respectability was also a sign of resistance, a claim by which female millworkers distinguished themselves from those managers and foremen who were "no better than they ought to have been."

The striking female millworkers who were "respectable married women" (and widows) came to their respectability by way of marriage rather than waged work. Decency was a virtue women felt they brought from home to the mill, a trait not incompatible with, but not indigenous to, factory life. The honour and dignity that they fought the strike in order to defend was not forged on the shop floor, but it was challenged there. Women took satisfaction in their work skills but drew their worth from their domestic roles. Although the community had long been accepting of working wives, in the crisis of the strike married respectability became rather like a modern Canadian pension: its portability was placed in question. Could a woman bring to militance the presence she held from her role as wife and mother? Would her authority, exercised by reference to domestic hierarchies rather than workplace relations, be acknowledged as relevant to the matters at issue in the strike? Once feelings had become heated, women on both sides of the dispute challenged the virtue of wage-earning mothers (a common enough strategy in the mainstream culture and in communities organized around the male breadwinner wage) to discredit their opponents. A vigorous supporter of the strike, Florrie Horsfall, who signed her letter to the editor "(Mrs) E. Horsfall," berated Elsie Smith for working while her children were young: "I came to Paris 28 years ago during the depression ... My family grew to four in nine years ... I had to go out to work to make ends meet, and found it very hard to leave my children. To Mrs Smith I would say that as she is the mother of six children she should stay home and do justice to her family" (*Brantford Expositor*, 3 March 1949). Elsie Smith in turn took on Grace Hockin, the strike supporter who could claim for her family thirty years' experience in and out of the mills. Hockin was doing Penman's work on a machine in her home before the strike, according to Smith, "able to make extra money, and still look after her home and babies without having to pay someone else to look after them while she went to the mill. After the little ones were tucked in for the night, and on Saturdays, she could make extra dollars, whereas we in the mill had regular hours ... I think Mrs. Hockin was very fortunate ... and should thank the company instead of running it down" (*Brantford Expositor*, 9 March 1949). Once there was an audience outside the community, once respectable women lent their standing as mothers to workplace militance, then the local, implicit exception that was protecting the honour of wives who took jobs in the mill was contested. The genie was out of the bottle. All the ways in which domestic and workplace roles, gender and class identifications might erode *or* reinforce one another came into play.

The reinterpretation of local events through the perspective of the mainstream culture masked the origins and muted the force of womanly militance. On the second day of the strike, the first Ontario Provincial Police officers arrived in town, at the invitation of municipal council and after a month's pressure by the firm (*Globe and Mail*, 20 January 1949; Parent-Rowley Collection, 11/6).[15] The next morning, in what the Toronto press described as a "wild melée" outside the hosiery mill, OPP officers arrested Gertrude Williams, 39, and Margaret Higgins, 18, and charged them with disorderly conduct. William Stewart, the UTWA field representative in Paris, summoned his best patriarchal indignation, portraying Williams and Higgins as vulnerable creatures by comparison with the real criminals at loose in the province: "I hate their guts, every one of them. Why aren't they out looking for Mickey MacDonald? Why aren't they out looking for him instead of arresting the wife of Charlie Williams, and shoving little girls around on a picket line?" *(Toronto Star*, 21 January 1949). Mrs Williams, in her own account, was less willing to be dismissed from the fray by virtue of her gender: "I was on the picket line not doing anything. There was some pushing and the next thing I knew I was arrested. But

I took some arresting. It took three of them to put me in a car. They tried to scare me and say my place was at home. I told them my place is where I want to put it. My place is on the picket line, and that's where I'll be, every day until its over" (*Brantford Expositor*, 22 January 1949). Mrs Williams was the odd-one-out in this exchange. The male union organizer believed she would not, and the provincial policemen believed that she should not, behave on the line in a way that would make her an effective picket. Mrs Williams alone among them seemed to credit her capacity for militance.

Horsfall, Hockin, Smith, and Williams had all worked for Penman's in the past and probably expected to be on the mills' payrolls at some time again in the future, though none of them was a Penman's employee late in 1948. Women in the community commonly moved in and out of the labour force as the needs of their households changed, and throughout their adult lives they construed their own well-being as intimately connected with conditions in the mills. Those neighbours and kin who shared child-care responsibilities and fashioned collective living arrangements in order to make waged work possible were knowledgeable and concerned and considered themselves parties to the dispute, even though they were not formally parties in the wage relationship (Parr 1987a). In Paris, as in the Lawrence strike of 1912 (Cameron 1985, 44–66), the female militance was forged and sustained in family and neighbourhood relationships. For women it was within these relationships rather than through union organization that the need for change in the workplace was most compellingly articulated and the most formidable alliances both for and against the strike were lodged. Friends and relations swelled the numbers of militants and widened their intelligence networks, their sources of sober counsel, and their tactical resources. Among the most dauntless of the pickets was the wife of a man who, once the boss in the shipping room and "a real good guy to work for" (Interview, James Baker), had been fired by a universally detested manager without apparent cause. When Eleanor Barrett, aged 22, was "plucked" from the picket line and charged with intimidation, her mother, who kept house for a family of five millworkers, immediately and publicly came forward to replace her (*Paris [Ontario] Star*, 3 February 1949; *Globe and Mail*, 3 February 1949). The day-care arrangements among neighbours and kin that had allowed mothers to hold down jobs in the mill now freed them to take their place on the line (*Toronto Star*, 21 January 1949).

Women controlled a considerable amount of housing stock in the town.[16] More than a third of Penman's female employees in 1948 owned their own homes. Boardinghouse keepers often ran their businesses out of more than one dwelling, (Historical Perspectives

Collection, interview, Hilda Scott Sharp), and older women workers bought small rental properties as a security against the days when they could no longer go to the mill. The homeowners were from those old mill families most strongly committed to the strike and, as a result, workers arriving in town to take jobs during the dispute found few boarding places open to them. Mrs Clem Smith, declaring that none of her properties would shelter "scabs" (*Globe and Mail*, 22 February 1949), evicted the tenants who had let the rooms they had rented in one of her houses to four Nova Scotian strike-breakers. The Nova Scotians had scuffled with her nieces and nephews on the line, and one of them had beaten up her nephew William in a Friday night disturbance on Main Street (*Brantford Expositor*, 22 February 1949). Several women whose back gardens opened onto lanes near the mills kept their kitchens open through the night, dispensing hot coffee and encouragement. Florence Lewis suffered badly later in the community for showing her loyalty to the strikers in this way, but Robert Fletcher, who, from his car, kept the night vigil on the Willow Street gate, remembered her as an "awful good woman" who had been more effective than any of the unionists from nearby centres in supporting the strike (Interviews, Robert Fletcher, Florence Lewis).

Family thus could be "the institution *par excellence*" by which women combined "to defend their collective interest," and "community based, female ties and networks" could "be mobilised to produce forms of female control." But the connection between gender and community solidarities and class-consciousness was contingent and volatile (Grieco and Whipp 1986, 120–1; Lamphere 1985, 521, 539; Santos 1985). Florence Lewis's elderly neighbour, who worked at Penman's until she was well on in her eighties, felt Lewis had betrayed both her and the accommodating employer who had given her work; they never spoke again. After a sideman in the Baptist church refused to walk the aisle with her to a pew, Lewis never thereafter attended a service (Interviews, Jean Hubbard, Florence Lewis).

The choice among family, community, and class interest that caused Charles Harrison and the ten other key union members to leave the UTWA in the wake of the strike vote was posed forcefully for the Pike and England families in the first week of the strike. On Thursday morning Doreen Pike, a finisher in the sweater mill, and her husband Arthur, a Department of Highways employee, were on the picket line. Doreen's mother would have been on the line as well, she declared later, but it was her turn to look after the children. Arthur Pike was arrested and charged by one of the thirty-seven Ontario Provincial Policemen now stationed in Paris at the request of Mayor William England. England was Doreen's father, Arthur's father-in-law. By

Thursday evening there had been ten arrests. Those released were critical of the conditions under which they had been confined. Three hundred strike supporters marched through the streets to Mayor England's house, calling out that the police were "outsiders" and carrying a petition protesting the presence of the provincials in town. There was shouting, singing, and a scuffle on the Mayor's front lawn. The police intervened. The next day it was rumoured that the Mayor would resign; he did not, but Doreen Pike withdrew from the picket line, declaring: "There was no excuse for them to make all that noise and disturbance." Soon afterwards her father was hospitalized in a state of nervous collapse, and he remained out of the action for the duration of the dispute (*Toronto Star*, 21 January 1949; *Globe and Mail*, 21 January 1949; *Brantford Expositor*, 21 January 1949; *Paris Star*, 27 January 1949). Ten days later, Charles Alexander, the long-suffering president of Local 153, also left the picket, claiming illness in his family, but also that he had "wanted no violence and this had led him to stay off the lines." The majority on council began to talk of the dispute as a "'family quarrel' in which the council should take no part" (*Brantford Expositor*, 8 February 1949). Some unionists and nonunionists alike came to use the imagery of family as a justification for retreat rather than engagement.

Like the spring floods that periodically tore through mills and millworkers' housing on the river flats, the strike was a ferocious and fearsome physical presence in town. The struggle to think about social relations within the community in a different way; to understand, as Charles Alexander wrote on 1 March 1947 to Val Bjarnason (Parent-Rowley Collection, 2/10) of the UTWA, "the forces underlying the formation of a social system and the transmutation of individuals and families to that system"; to act on the basis of that understanding to claim rights, name abuses, and acknowledge the divide that lay between workers and bosses in town and finally choose sides – all of this was arduous, but not alien. There were comforting similarities between the promises of a fair hearing and a square deal in an organized workplace and honoured family, neighbourhood, and community values. No one had ever behaved in public in the ways that the script of the strike required. Parades had always been cheerful national or seasonal celebrations in commom cause, not angry assertions of claims or demonstrations of force. The mill gates had been places to exchange gossip and cigarettes, not curses and blows. On the first days, as the picket lines went up, a nonstriker told local historian Donald Smith (1980, 60) that many who went into the mill "felt sick at heart" and went out at night to see that "those fellows out in the cold got some coffee and sandwiches." But by midweek there were scores of provincial policemen, squads of big-city reporters, and several union organizers in town, giving "the impression of a small invasion from some other part of the country" (*Toronto Star*, 21 January 1949). These groups were defining the limits of the stage, and simple neighbourly gestures that were incompatible with the actors' strictly scripted parts were squeezed off into the wings. Many in the community remained heartsick, but, uneasily, they played out their new roles. The pickets on the graveyard shift went for their coffee to Mrs. Lewis, Mrs. Corrigan, and the other women who kept their kitchen lights burning through the night.

The picket line was to provide a barrier between intending nonstrikers and their workplaces in the mill. Its devices included silent shaming, verbal intimidation, and physical force. Of these, only shaming was considered a womanly act, but a picket that functioned using shaming alone would have won no public attention and would have served only to discomfit, but not inhibit, workers passing through the line. The unaccustomed rhetoric of the barroom and postures of the brawler both limited and enhanced women's effectivness as picketers as well as both the way they were perceived and the way they felt about themselves. The resort to violence could compromise the acknowledged respectability upon which their claims to authority had been based or, by the profoundness of its anomaly, accentuate the urgency and the justice of their cause. The sensation of being out of bounds could be exhilarating and empowering, drawing women back to the picket again and again to savour its exotic pleasures; it could also cause them to retreat from its very possibility or, having glimpsed its fascinations, to draw back in self-revulsion (Hall 1986, 354–82; Mason 1987, 145, 147, 150; Costello 1987, 291, 299; Strom 1983, 360, 364, 370; Nash 1975, 261, 270).

Mildred Hopper (Interview), who was early convinced that, with the union, Penman's could be made a better place to work, went home to her parents' farm in Princeton the morning the strike began and stayed there until the strike was over: "I went out but I didn't want to be any part of it. I think that [being on the picket line] would be worst of all." Lottie Keen (Interview), a widow who was one of only two women in the looping department to honour the picket, could not bring herself to join the line. "Some of them [the women picketers] were very outspoken, like they'd follow you and call you scab and all this sort of stuff ... so I didn't feel bad about it, but it did make a lot of ... [voice tailing off]." May Phillips (Interview) remembered most of the picketers as being women: "Most of them were women that worked there you see ... When that started I just stayed home. I lived in Brantford then, so I didn't see any sense in travelling up here and getting into a hassel. I don't like that kind of thing anyway ... Getting into scraps all the time." Her husband was a member of the auto-workers' union in Brantford, but he distrusted the politics of the UTWA and she deferred to his judgment. "My husband says 'you're not going up there while they are on strike' because he knew what that union was like." Frances Randall's husband (Interview, Frances Randall) stayed out in respect for friends who had joined the union, but he gave clear orders to his wife, "don't you dare go over there," and she didn't. Within the union movement, even among female activists, it was difficult to acknowledge and defend womanly militance on its own merits. Jessie Bragg, five years a member of the UAW-CIO in Brantford, had helped the UTWA recruit members in Paris and had spoken regu-larly at their meetings; but as the clashes on the picket line grew more violent and sympathy for the strikers appeared to wane, she felt compelled to reconfigure the dispute for the press: "In looking over the picket lines, I have noticed at least 90 percent of the pickets are veterans from the last two wars and some veterans of both wars. We, as war workers ... were sure that those boys had the best during the war, so why don't we, as citizens of this town, whether it be the town fathers, Board of Trade, merchants or ministers of all denominations, get behind this affair and see that these boys have the best now?" (*Brantford Expositor*, 14 March 1949).

Women's postures on the picket line were sharp-edged collages of their old and new roles, compelling for both their dissonance and their familiarity. In the first week of the strike, "when the provincials got out of their cars and stationed themselves by the gates of the three mills, they smiled at the pickets and joked with some of the girls. The pickets found themselves responding in kind. The cops were pleasant." "Here come the other pickets," a woman exclaimed, "there's more in their picket line than in ours" (*Toronto Star*, 20 January 1949; *Brantford Expositor*, 21 January 1949). In the second week, when the weather turned bitterly cold, women "far outnumbered" men on the line; "they said the weather did not bother them" (*Brantford Expositor*, 26 January 1949). Clara Farr remembered feeling queenly on the picket. "Oh that was something! I never owned a fur coat in my life and somebody was good enough to bring me a fur coat so that I could go on the picket line"; "My girlfriends and I really had a lot of fun. It brought us a lot closer together. We became close friends"; "We had dances; we had parties" (Historical Perspectives Collection, interview, Clara Farr). It was a nonstriker who recalled for Donald Smith (1980, 55), with barely submerged glee, that "[o]ne girl was very mouthy. She yelled all kinds of things at workers who were entering the mill. She even yelled them at a relative of hers. Finally the sergeant in charge of the police at that gate said to one of the relatives, 'The next time she yells at you, I'll turn by back and you slap her face.' Some of the women used hatpins, even jabbed the police in the rump. They carried them in their purses. They'd stick them into the cops when the cops were in line ... One cop had to go to the hospital for an infection in his rump. He was there for a few days. The stabbing was worse at night when the light was poor." Police, journalists, and pickets alike depicted female militance with the imagery of feminine wiles.

Gender sensibilities could also mute the force of the conflict, however. On the line males who cursed females did no credit to their side; men who claimed injury from women's sharp tongues were objects of mirth; shouting matches between women neighbours were construed as catfights in the press (Interview, Ida Glass; *Paris Star*, 17 March 1949; *Brantford Expositor*, 25 February 1949; *Toronto Star*, 24 February 1949). Gendered identities were masks that changed in the shifting light and shadow of the dispute; they were mercurial, unpredictable in their effects upon public sympathy.

On 19 January Elizabeth May Cardy, who had run one of the large flatbed, full-fashioned knitting machines in the hosiery mill since 1929, was injured in an altercation, allegedly with Arthur Gignac and Lillian Gillow, as she tried to cross the line. Nine days later she suffered a stroke. At seven-thirty on the morning of 3 February, charges of intimidation with violence were laid against Gillow and Gignac in connection with the incident. The pickets and police were massed that morning at the Elm Street gate. Gillow was home with the flu as Gignac, unaccountably, it seemed to the other pickets, was dragged off the line into a squad car; what the press called a "free-for-all" broke out. Hendrika Bethune, a twenty-two-year-old winder, was forced to the ground in the centre of a crowd; Robert Williams, whose mother had been arrested early in the dispute, thought Bethune hurt and as he went to her aid was grabbed by a policeman. Helen Murphy, the secretary-treasurer of the local (and, at well over 200 pounds, a formidable presence on the line), waded in to protect young Williams and move the officer aside. There were nine arrests in the twenty minutes before the eight o'clock whistle blew; Murphy was charged with assaulting a policeman. Just after eight, the news came that Elizabeth Cardy had died. That day's front-page news showed Bethune writhing on the ground, surrounded by police. The next day a hundred pickets massed at the same gate and there were four arrests. Leta Morrison, aged thirty-seven, was run over by a car filled with non-strikers and taken to hospital. The newspapers showed Gladys Burtch, a twenty-three-year-old finisher, being carried from the scene in the arms of her father and Val Bjarnason from the UTWA after, the pickets claimed, she had been kicked in the stomach by a policeman (*Brantford Expositor*, 3 and 4 February 1949; Parent-Rowley Collection, 2/11).[17]

The arrests of 4 February brought the total in the first three weeks of the strike to twenty-seven. In the two months that followed, until the strike was settled on 9 April, the police laid seven more charges – five against UTWA organizers, and only two against town residents. Though Coroner W.J. Deadman's report released on 11 February determined that Cardy's death was not due to violence (she suffered from Bright's disease and twice previously had stopped work for extended periods [*Brantford Expositor*, 11 February 1949, 16 February 1949]) and though Morrison and Burtch recovered from their injuries, the will to make the picket a physical barrier around the mills was gone after the incidents. The 80 to 100 workers who still marched the picket lines substituted songs and snake dances for curses and blows. Coverage of the strike now concerned events in the courts rather than the street.

The trials began 7 February. The court list, with its charges against twenty-three persons, was unprecedented in Paris. By the time the accused "and police and other witnesses have been admitted to the small Council Chamber," the press predicted, "there will be scarcely a seat available for anyone in the role of mere spectator" (*Brantford Expositor*, 7 February 1949). The standard charge laid, against male and female picketers alike, was creating a disturbance and intimidation with violence. Police testimony was commanding in the court. The rights of pickets were narrowly construed in the law. Three out of every four charges laid against strikers resulted in convictions (Parent-Rowley Collection, 11/7). But Magistrate R.J. Gillen of Brantford viewed the witnesses and judged the behaviour required for a conviction through gendered expectations. In their questioning, both the crown attorneys and the defense counsel used presuppositions about what manly and womanly conduct would have been like in order to elicit witness testimony about what had happened on the line. Neither strikers nor nonstrikers had set aside their identities as women and men when they assumed their new roles on the picket line; they both recalled their familiar, gendered reflexes to recoil and defend and listened as their unaccustomed postures of militance, which were so electrifying to claim and display, were made mild and conventional in legal summation. Looking out into the crowded council chamber, knowing their tale would be told and retold, they refashioned their recollections so that they became aligned with those of their sex.

The reaction of the participants rather than their acts was key for Magistrate Gillen in ruling upon intimidation. The pattern in his judgments became clear in the first day of the trials. A male picket captain was accused of intimidating Gordon Parsons, a knitter-mechanic crossing the line; Parsons testifed that the picket had said, "You're bloody brave while the cops are here, but I'll get you yet, you —," but insisted that the threat had not made him fearful. Dismissing the charge, Gillen said that "as Parsons was not afraid of the accused, there was no intimidation." The requirement that the persons who were intimidated acknowledge before the court that they had been made afraid, when the avowal empowered their adversary and revealed their own vulnerability, shaped the court proceedings according to patriarchal and other more proximate, political relations. Older women would agree that they had feared younger women; thus eighteen-year-old Margaret Higgins did intimidate fifty-year-old Rose Lewis. A woman threatened by a man would admit fear. Some older men would acknowledge fearing younger men. But no man would own up to being frightened by any woman on the line, no matter how formidable her presence – with one exception: the crown attorney, by implying that a union organizer – whether male or female – was a communist, could secure an admission of fear and a conviction against the accused from any witness before him (*Brantford Expositor*, 8, 9, and 15 February 1949).

Men's actions to protect women or the elderly of either gender were viewed favourably by the court; but for those who had been in the vicinity of violence, maleness alone could carry the implication of guilt. Lillian Gillow was a thirty-three-year-old widow, a finisher from the hosiery mill, living with her two children in her widowed father's house. She was charged with intimidation with violence after Mrs Cardy suffered her stroke. Gillow testified that she had been alone, holding Mrs Cardy's arm and attempting to talk to her, when two men, John Rogers and Martin Hogan, tried to help Cardy up the steps to a high loading platform outside the mill. Gillow said Arthur Gignac, the twenty-four-year-old knitter also accused in the incident, was not nearby. Gignac testified he was several yards away, and defense witnesses concurred that he had not touched Mrs Cardy. Rogers and Hogan, however, claimed that, looking back, they had seen Gignac grab Cardy's right hand. The court had heard that Cardy had been in ill health for two years but, convicting both Gillow and Gignac, Magistrate Gillan reasoned that "there could be no doubt that Mr Hogan and Mr Rogers were trying to help Mrs Cardy up to the platform and it would not have been necessary for them to help her, unless she were being held back. He believed that Arthur Gignac was holding her back and Mrs Gillow was pushing her" (*Brantford Expositor*, 16 February 1949). In view of the conflicting testimony about

Gignac's location, the magistrate based his judgment on his appraisal of what force a woman might be able to exert and, even though Cardy was frail and the loading dock high, determined that Lillian Gillow alone could not have dislodged her footing.

If men faced a more difficult time at the hands of the courts, women suffered greater nonjudicial penalities as a result of their participation in the strike. The settlement that was agreed upon won only one concession from Penman's – that there would be no discrimination in rehiring – but this clause left the firm wide powers to determine the availability and suitability of candidates for jobs in the mill (Parent-Rowley Collection, 11/2).[18] Women's longer experience with conditions in the mills and their more limited opportunities elsewhere in the labour force made change at Penman's an urgent priority for them; yet older women, wives, and widows who were exceptional in their militance were left most vulnerable in the wake of the strike. Women made sense of their workplace grievances in different ways and put different words to their militance than did men. They formed solidarities and defended them differently. Contemporary unions did not comprehend these differences well; nor did they find ways to amplify this distinctive force in women's activism in order to forward their workplace struggles. Organizers, as well as the press, the officers of the court, and the police, were conceptually constrained to feature female unionists as womanly. They thus undercut the force of women's activism, editing it out of the power relations of the strike. Women strikers themselves simultaneously owned and disowned their militant postures, particularly retrospectively, and, in a community where union organization had failed, they were unable to frame a conception of themselves that would contain this conflicting and dissonant experience. In this dilemma they were not, and are not, alone.

NOTES

1 Lists of strikers and nonstrikers prepared from the formal interviews and informal discussions with townspeople during my eight months in town were accurate, without exception, when compared with the personnel records of the firm. Each person interviewed has been assigned a pseudonym, and all references are to pseudonyms.

2 Cited documents in the Parent-Rowley Collection: 1/15, "History of UTWA Organization"; 5/5, "The Canadian textile industry, an economic review," 8 November 1949.

3 Sources of the information include: Parent-Rowley Collection 2/10; 2/13; 11/5; 17/2; Parent Collection, Regan-Rowley Correspondence,

25 March, 19 July, 4 August 1946; *Paris (Ontario) Star*, 12 September
1946; Canadian Labor Congress, Collection, Parent to Sullivan,
4 December 1946.

4 Parent-Rowley, 2/10; Alexander to Rowley, 23 March 1947; see also
UTWA *Canadian District News*, 9 May 1947.

5 Brockbank, Hogan, and Smith were interviewed on this topic for a
local history project conducted in Paris in the 1970s. Transcripts of the
Smith and Brockbank interviews are in the Paris Public Library in the
Historical Perspectives Collection, while the tape of the Hogan interview, which was not transcribed, is with the records of the project at
the Paris Town Hall.

6 Signed cards were submitted from 452 of the 554 employees who were
considered members of the bargaining unit. When the vote was held on
April 2, in a unit defined to include 649 people, there were 328 votes
for the UTWA and 240 for the company union, a margin of 3 votes
above the level required for certification (*Paris [Ontario] Star*, 25
March 1949; *Brantford Expositor*, 3 April 1949). Over the next three
months representatives of the union and the firm met six times but
made no major progress. A provincially appointed conciliator failed to
effect a settlement in August.

7 The union had asked for a 20 cent hourly increase: 5 cents for the cost
of living, and 15 cents to bring Penman's wages closer to average rates
paid in the district. Citing the continuing activity of the company union
group, they also asked for a maintenance-of-membership clause, requiring members to remain in good standing within the union for the duration of the contract. For a history of the negotiations from the union
perspective, correspondence concerning the constitution of the conciliation board, union and company submissions to the conciliation board,
and the majority report, see UTWA files, 11/6; for the petition for conciliation and both the majority and minority reports, see Company and
Strike Files, 1948, P–S. The report on Penman's response to the concila-
tion is in the *Brantford Expositor*, 25 November 1949.

8 Document cited (Parent-Rowley 11/6): Stewart to Wren, 27 October
1948.

9 The size of the Penman's payroll has been inferred from the personnel
records of the firm, which list by day, month, and year the periods of
employment for each worker. The wage books themselves were lost in a
flood. The personnel records show the cause for each interruption of
employment and explicitly identify the strikers. These records are in the
Penman's Archives, Cambridge, Ontario. These is only one, undated,
membership list in the UTWA files for 1948 (11/7). It includes 433
names, 19 fewer than the number of cards filed in the March 1948 petition for certification. The hourly-rated payroll would be larger than the

bargaining unit by the number of supervisors, who were preponderantly male; many male maintenance workers crossed the line by agreement with the strike committee and thus do not appear in the personnel records as strikers, although they supported the work action. Both of these factors would tend to understate the proportion of men who supported the strike.

10 The household structures of 256 of the women who worked in Penman's four Paris mills were determined by linking the personnel records in the Penman's Archives with the municipal assessment rolls in the Town Hall. Twenty-six per cent of these women, 67 of the 256, lived in female-headed households. Irene Cobbett, who lived with her widowed mother and sisters, describes the evening visits from "union men" (Interview, Irene Cobbett).

11 For example in the largest mill, the hosiery mill, in 1948 the average age of women workers was 32, the average for women strikers 36; the average male striker was 30 years old, the average male waged employee was 29. At the same mill, 45 per cent of male strikers and 29 per cent of male employees were single. Eighteen per cent of women strikers and 8 per cent of women employees were widows; 64 per cent of women strikers and 55 per cent of women employees were wives.

12 The quotation is from a letter dated 6 December 1948 to Helen Murphy, the secretary of Local 153. It was signed by ten senior employees of the firm (eight men and two women), including five fixers and two experienced knitters, who withdrew their membership from the union in the wake of the strike vote on 22 November.

13 See also the *Brantford Expositor*, 22 February, 5 March, 12 March 1949.

14 The strike vote was taken 22 November 1948 (*Brantford Expositor*, 23 November 1948). By this time the voluntary irrevocable check-off was a common feature of UTWA contracts in Ontario. See UTWA files, 5f3. For the statement that the strike was primarily about union security, see, from the union side, the letter from Eugene Stratton, a brillant fixer from the hosiery mill, *Brantford Expositor*, 16 March 1949; and for the management side, D.L.G. Jones, solicitor for Penman's, *Brantford Expositor*, 25 January 1949.

15 Parent-Rowley, 11/6: William Haggett, clerk and treasurer, Town of Paris, to Helen Murphy, secretary of Local 153, UTWA.

16 Thirty-seven per cent of the female employees in the hosiery mill and 36 per cent of the female employees in the sweater, yarn, and underwear mills owned their own homes in 1948. See the discussion of boardinghouse keeping and real estate investment among female midlands emigrants in Parr 1987b. The notation "could not get a boarding house" was recorded on the personnel cards of strike-breakers (Personnel Records, Penman's Archives).

17 See also the Strikes and Lockouts Files; Parent-Rowley 2/11; UTWA file, notes on court sessions.
18 Parent-Rowley, 11/2: Memorandum of settlement, 9 April 1949; *Brantford Expositor,* 9 April, 11 April 1949.

REFERENCES

Benenson, Harold. 1985. "The community and family bases of U.S. working class protest, 1880–1920: a critique of the 'skill degradation' and 'ecological' perspectives." *Research in Social Movements, Conflict and Change* 8: 109–32.

Cameron, A. 1985. "Bread and roses revisited: women's culture and working-class activism in the Lawrence strike of 1912." In *Women, Work and Protest: a Century of Women's Labour History,* R. Milkman, ed, 42–61. Boston: Routledge.

Canadian Labor Congress Collection. National Archives of Canada, Ottawa.

Company and Strike Files. Labour Department Conciliation Services Branch. RG7 V-1-b 31. Archives of Ontario.

Costello, C. 1987. "Working women's consciousness: traditional or oppositional?" In *"To Toil the Livelong Day": American Women at Work, 1780–1980,* C. Groneman and M. Norton, eds, 284–302. Ithaca: Cornell University Press.

Frankel, L. 1984. "Southern textile women: generations of survival and struggle." In *My Troubles are Going to have Trouble with Me: Everyday Triumphs of Women Workers,* K. Brodkin and D. Remy, eds, 39–60. New Brunswick, NJ: Rutgers University Press.

Grieco, M. and R. Whipp. 1986. "Women and the workplace: gender and control in the labour process." In *Gender and the Labour Process,* D. Knights and H. Willmott, eds, 117–39. Aldershot: Gower.

Hall, Jacquelyn Dowd. 1986. "Disorderly women: gender and labor militancy in the Appalachian South." *Journal of American History* 73. 2: 354–82.

Historical Perspectives Collection. Interviews. Paris Public Library and Paris Town Hall.

Interviews. 1984–85. Paris Industrial History Project Collection. Queen's University Archives, Kingston, Ontario.

Jameson, E. 1977. "Imperfect unions: class and gender in Cripple Creek, 1894–1904." In *Class, Sex and the Woman Worker,* M. Cantor and B. Lurie, eds, 166–202. Westport: Greenwood.

Jones, B. 1984. "Race, sex and class: black female tobacco workers in Durham, North Carolina, 1920–1940, and the development of female consciousness." *Feminist Studies* 10. 3: 441–51.

Kessler-Harris, A. 1985. "Problems of coalition-building: women and trade unions in the 1920s." In *Women, Work and Protest, a Century of Women's Labour History,* R. Milkman, ed, 110–38. Boston: Routledge.

– 1986. "Independence and virtue in the lives of wage-earning women: the United States, 1870–1930." In *Women in Culture and Politics*, J. Friedlander et al., eds, 3–17. Bloomington: Indiana University Press.

Lamphere, L. 1985. "Bringing the family to work: women's culture on the shop floor." *Feminist Studies* 11. 3: 519–40.

Mason, K. 1987. "Feeling the pinch: the Kalamazoo corsetmakers' strike of 1912." In *"To Toil the Livelong Day": American Women at Work, 1780–1980*, C. Groneman and M. Norton, eds, 141–60. Ithaca: Cornell University Press.

Nash, J. 1975. "Resistance as protest: women in the struggle of Bolivian tin-mining communities." In *Women Cross-culturising Cross-culturally*, R. Rohrlich-Leavitt, ed, 261–71. The Hague: Mouton.

Parent Collection. Private collection of Madeline Parent, Montreal.

Parent-Rowley Collection. MG31 B19. National Archives of Canada, Ottawa.

Parr, J. 1987a. "Rethinking work and kinship in a Canadian hosiery town, 1910–1950." *Feminist Studies* 13. 1: 137–62.

– 1987b. "The skilled emigrant and her kin: gender, culture and labour recruitment." *Canadian Historical Review* 68. 4: 529–51.

Personnel Records. Penman's Archives, Cambridge, Ontario.

Ross, E. 1985 "'Not the sort that would sit on the doorstep': respectability in pre-World War I London neighbourhoods." *International Labour and Working Class History* 27: 39–59.

Santos, Michael. 1985. "Community and communism: the 1928 New Bedford textile strike." *Labour History* 26. 2: 230–49.

Smith, Donald. 1980. The Penman's strike of 1949. Typescript. Historical Perspectives Collection, Paris Public Library.

Strikes and Lockouts Files. Department of Labour. National Archives of Canada, Ottawa.

Strom, S. 1983. "Challenging 'women's place': feminism, the left and industrial unionism in the 1930s." *Feminist Studies* 9. 2: 359–86.

Tilly, L. 1981. "Paths of proletarianization: organization of production, sexual division of labor, and women's collective action." *Signs* 7. 2: 400–17.

Turbin, Carol. 1987. "Beyond conventional wisdom: women's wage work, household economic contribution and labor activism in a mid-nineteenth-century working-class community." In *"To Toil the Livelong Day": America's Women at Work, 1780–1980*, C. Groneman and M. Norton, eds, 47–67. Ithaca: Cornell University Press.

UTWA files. MG31 B19. National Archives of Canada, Ottawa.

Waldinger, Roger. 1985. "Another look at the International Ladies' Garment Workers' Union." In *Women, Work and Protest: A Century of Women's Labor History*, R. Milkman, ed, 86–109. Boston: Routledge.

5 Women, Work, and Place: The Canadian Context

SYLVIA GOLD

Women's participation in the Canadian paid labour force has increased significantly since the beginning of this century. This increase would not be represented by a graph line showing steady increase across the decades but by a line broken by peaks and valleys while persistently moving upwards. These peaks and valleys can be explained by social, economic, and political events and processes of the corresponding era. Attitudes towards women in paid work, the availability of labour employment rates, the state of the economy, and wars have had great impact on women's employment. Events such as International Women's Day (IWD), which is held on 8 March each year, grew out of concerns about women's labour-force participation and women's votes; they attempted to voice and represent the life situation of women and to search for solutions to oppressive working conditions and the powerlessness of women. Discussion will focus primarily on four issues that relate directly to women's labour-force participation: 1) mobilization of women in the early 1900s that led to annual celebrations of International Women's Day; 2) a Canadian women's participation in the waged labour force; 3) women and entrepreneurism; and 4) the social structures, values, and attitudes that either inhibit or facilitate women's participation in the paid work force. These subjects continue to motivate women's organizations in Canada today, since changes in economics and social policies have not caught up with the experiences and aspirations of most women.

WOMEN CELEBRATING

Slogans on T-shirts often capture and express currents of change in social values and attitudes. In a few words, just like advertising commercials, they play upon emotions, frustrations, desires, and, often, controversial if not taboo subjects. One slogan that comes to mind in this discussion of women, work, and place first appeared in the late 1970s: A WOMAN'S PLACE ... IS IN THE HOUSE OF COMMONS, which evokes the old saying, "A woman's place is in the home [in the kitchen, barefoot and pregnant!]" and juxtaposes it with the image not of the house but of the House, where the laws of the land are debated and rejected or accepted, and where the representatives of the people – all of the people – speak, decide, and lead us into the future. And yet, after the Canadian federal election of 1988, only 13 per cent of federal Members of Parliament were women. This figure is higher than that for both Great Britain and the United States (about 6 per cent) in the same year, but falls far below the situation in Norway, where 40 per cent of the House of elected representatives that year were women (Hawke 1988).

The slogan suggests several themes, among which are a recognition of women's activities outside the home and the role for women in developing the legal and political framework for economic activity. The significance of the disproportionately low number of women in the political decision-making arena is that the experiences and aspirations of women are largely absent in a male-norm-centred environment. It is an overwhelming challenge to pass legislation that is satisfactory to women on child care or pensions in an environment where the large majority have never coped with the day-to-day demands of child rearing, or the anxiety of finding reliable child care, and have little comprehension of reaching retirement without having had the opportunity to accumulate work-related pensions or similar retirement income.

Thus, *Women, work, and place* has a distinct meaning in the present. Statistics demonstrate the rapidly changing patterns of women's lives and their movement into nontraditional places, like the legislature, automobile assembly lines, and the science labs. Our legislation and values are changing in response to women's new sense of place and, equally important, to society's dependence on this evolution. Historians, sociologists, and geographers will look back two or three decades from now and separate the threads of events, currents of thought, and activities to explain the situations we are living through, and they will speak of labour-market sectoral shifts that call for more skilled workers in some areas and fewer in others; note will be made of changing patterns of daily, weekly, and yearly work; problems of family adaptation to the demands on parents will be described and analyzed; the role of the women's movement – or perhaps women's movements – will be identified as ever present threads running through all other analyses. The current struggle for recognition of the work that women do in the home to maintain the well-being of all family members will, no doubt, be more sympathetically reviewed twenty years from now.

The major impetus for change in the lives of women today is the necessity for their economic self-sufficiency. For example the high divorce rate has increased the incidence of families headed by women; there is a higher rate of poverty among single mothers and elderly women compared to men; and there is the increasing desire of women to make their own decisions about their lives. In the quest for equality, the ability of women to ensure an adequate income for themselves and their families has become of primordial importance. While not all women actually belong to groups that identify with the modern women's movement, most can identify with its goals, such as securing the rights of homemakers to adequate pension, equal pay for work of equal value, access to reproductive health care and abortion as medical services, accessible, affordable, quality child care, and community supports for families (however families are defined). Women's groups today no longer accept the idea of "everything in its place," but argue that no place can be off limits to women.

Yet a fascinating history of IWD by Renée Côté (1986) reminds us that the more things change, the more they seem to stay the same. The IWD was instituted in 1910 at the second International Conference of Socialist Women in Copenhagen, following the successful American Women's Day celebrations in 1909 and 1910. Initially it was the American Socialist Party that, in July 1910, recommended to the International Conference of Socialist Women that the last Sunday of every February be designated International Women's Day. The IWD goal was to stimulate the participation of socialists in the struggle for women's right to vote. Women's suffrage emerged as the major theme of subsequent Women's Days.

On 27 February 1910, Women's Day was celebrated from coast to coast in the United States. Women spoke of economic and political equality, the vote, and their deplorable working conditions. Carrie Allen spoke of the slavery of women in factories and of women closed up in their kitchens. The strike of shirtwaist makers, which lasted from 22 November 1909 to 15 February 1910 and involved 20,000 to 30,000 workers – 80 per cent of whom were women – had just ended. This was the first massive strike by women against intolerable working conditions, and it was accompanied by widespread arrests and police brutality. It was not to be the last women's strike on the North American continent, however, as strikes and other job actions by nurses in the provinces of British Columbia, Alberta, Ontario, and Quebec in recent years have demonstrated. Ethel Whitehead makes the following points: first, more and more women are being forced to work in industries; secondly, while they have struggled with men for workplace rights, they also have had another battle (the "handicap of their sex"), because with rare exceptions they are paid less than men; thirdly, woman's sex becomes a commodity so that maternity, which should be her greatest joy and the consecration of her femininity, becomes a malediction (Côté 1986).

But in that first decade of the twentieth century, two major currents of thought emerged. One, utopian Christian (cooperative) socialism, favoured an autonomous women's movement; the other, scientific Marxist socialism, found the former too conservative and believed in class struggle rather than struggles between the sexes. The two currents of thought persisted, in conflict, and another notable conflict developed, too – that between those who argued for the vote for women and those who were working for improvements in the paid work force and were unwilling to contribute their energies to women's suffrage. Emma Goldman (1969, 204) maintained that all political regimes of her time were absurd and unresponsive to life's urgent problems. She criticized middle-class American women who saw themselves as equals of, or even superior to, men in their qualities and virtues, and she scoffed at these women who expected the vote to bring miracles. Elizabeth Gurley-Flynn of the Industrial Workers of the World said that the suffrage movement did not concern the working class. She saw the fight for women's rights as a class issue and saw no common interests among all women; there was no "war of the sexes, and no natural solidarity among women only" (Côté 1986, 122). The divisions between the women who sought the realization of socialist goals and the women fighting essentially for the vote were clear.

Since 1910 women in Canada, Germany, Sweden, France, and Russia, have been celebrating – or at least holding events – on International Women's Day. How has the fight for women's rights developed since that time? What is most provocative about the events of the first International Women's Day is that they were born out of women's articulations of their own experiences and aspirations. The tragedy is that this cauldron of activity was not powerful enough to catapult women into political decision-making power or to generate different directions for political and social thought and influence, directions that are sympathetic to the equality of women and women's integrity.

Despite the strength of recent feminist scholarship, existing social sciences such as economics, human geography, and sociology still offer poor foundations for the analysis of women's experiences. Assumptions, explicit or implicit, of women's dependence on men and men's responsibility for the care of women and families do not stand up under theoretical scrutiny, yet they continue to inform research in these sciences, and the literature they generate does influence Canadian legislation and policy. Until and unless we have responsive and reflective assumptions about women's roles as a foundation for new theories with which to influence new equality-based laws and programs, progress will be slow and frustrating.

To illustrate the point, if the principle of equality and recognition of everyone's need to contribute to, and benefit from, family were widely accepted, would we now be arguing about the cost of child care? Would employers expect upwardly mobile employees to spend unreasonably long hours at work and away from family to demonstrate their loyalty and productivity? And would one's concern for family be construed negatively by employers? The arguments brought forward in the early IWD activities were not sufficient to bring about equality for women. The subjects articulated then – the wage gap, working conditions, and family responsibility – remain important today. Despite the increasing participation of women in work outside the home, the issues of the proper place of women and the valuing of women's work remain urgent.

ECONOMIC PROGRESS TOWARD EQUALITY

The number of Canadian women in the waged labour force has increased dramatically since the early 1900s, when about 16 per cent of women were employed outside the home. In Canada, as elsewhere, these women spent ten to twelve hours per day, six days a week, in noisy, poorly ventilated factories or shops. Many, if not most, of these women were single or widowed who could not count on fathers, brothers, or sons to provide for them, either. In other cases the husbands of many married women simply could not earn enough to cover the family's expenses. Women's contribution to the formal economy continued to grow in the ensuing decades.

By 1970, 38 per cent of the women in Canada were working outside the home. That figure rose to 57 per cent by 1988, compared to an approximate 77 per cent participation rate for men; it has since fallen to a 53 per cent participation rate for women and, more dramatically, to 67 per cent for men in 1991. The absolute number of women in the labour force peaked at 5.8 million, compared to 7.4 million men, in 1988 (Statistics Canada 1988a); the reduction to 5.6 million and 6.8 million, respectively, in 1991 reflects the general decline in the work force. Nonetheless women's participation in the labour force has continued to increase proportionately, growing from 36 per cent in 1975 to 45 per cent in 1991 (Statistics Canada 1993 [spring], 3).

Since World War II more and more married women, both those with and those without children, have been entering the labour force. In 1983, for example, 49 per cent of women in the work force had children under the age of 3, 56 per cent had children under 5, and 62 per cent had children under 15 (CCLOW 1986). By 1989 single women had the highest participation rate (68.2 per cent), followed by married women (59.9 per cent) and widowed, separated, or divorced women (36.1 per cent). Between 1981 and 1989 the most notable increase was among women with employed husbands; their labour-force participation rate increased from 57.6 per cent to 71.2 per cent. In 1989 mothers whose youngest child was between the ages of 6 and 15 were more likely to be labour-force participants than were mothers with pre-school-aged children or mothers with children over 16 (Labour Canada 1991, 5). But this increase in the number of women in the work force was not marked by a parallel increase of women in male-dominated professions. Where women were 11.4 per cent of workers in male-dominated professions in 1971 they had increased to only 18.6 per cent of these categories by 1981, and to 23 per cent in 1986 (Statistics Canada, 1989 [spring], 15).

At the beginning of the century studies showed that women worked an average of 55 hours per week in the home. Despite advances in technology, recent data confirm that, in 1986, women at home full-time still work an average of 55 hours per week. It is also interesting to note that married women employed in the labour force work on average 35 hours a week in the home, in addition to the hours spent in the paid work force. Husbands of full-time homemakers, however, work only 10 hours per week in the home, while husbands of women in the paid labour force work 11 (CACSW 1987).

Although women are now acknowledged to be "full-fledged" members of the labour force, the situation of women in the labour market has not substantially improved in this century. The historic problems of a large wage gap and occupational segregation are still of grave concern. There is no doubt that the standard of living for both women and men has improved considerably since the 1900s. Yet it is also a fact that the wage gap has been fairly constant over the past fifteen years. In 1970, for example, women who worked full-time, year round were found to earn about 60 per cent of the wages of men who worked on the same basis. By 1980 there was only a 4 per cent improvement.

By 1991, women employed full-time, year round earned, on average, $26,800, just 70 per cent of the average earnings of comparable men; and nearly twice as many women as men had earnings below the poverty line (Statistics Canada 1993 [spring], 5).

Inequality between men and women workers does not end with wages. Women workers also tend to be employed in smaller, nonunionized, and less profitable industries, where wages are lower than in larger, unionized shops and where there are few, if any, benefits. Since unemployment insurance benefits, pensions, sick leave, maternity leave, and other benefits are often tied to wage levels, women not only receive lower wages than men when they are working, but they receive much less than that when they are bearing children or when they are old, sick, or unemployed. In 1989 women received lower Canada Pension Plan (CCP) benefits than men; they accounted for only 29.3 per cent of the total CCP recipients receiving monthly pensions of $200 or more (Labour Canada 1991, 1). The weekly benefit paid to women who qualify for maternity benefits under the unemployment insurance programme aptly illustrates the low earnings of women workers. In 1989, for instance, when the maximum possible weekly benefit was $363, women who claimed benefits received an average of $234 a week (Employment and Immigration Canada 1989). They are thus penalized all their lives, in that they earn less during their working years and are likely to have no pension in their retirement other than the public "Old Age Security" (Gee and Kimball 1987).

Women's work in the paid labour force has always tended to be concentrated in a narrow set of occupations. From the beginning of this century to 1971, 50 per cent of the female labour force was working in only six occupational categories: dressmakers and seamstresses, servants, nurses, teachers, office workers, and saleswomen and clerks (Marshall 1987). Except for nursing and teaching, all of these are still low-paying and low-status occupations, despite the recently voiced dissatisfaction with nurses' wages and working conditions, for example, which was expressed through militancy in 1988. We are also aware, however, of the barriers faced by women in these occupations, particularly their exclusion from decision- and policy-making, despite their numbers as practitioners. Today it is true that the number of occupations available in our economy has greatly expanded (the 1981 Census lists 500 occupations) but women are still segregated in twenty occupations, which are dominated by secretarial, clerical, and sales positions (Day 1987). In 1989, 57.4 per cent of the total female labour force was still concentrated in these occupations (Labour Canada 1991, 1).

In terms of occupational trends, a significant proportion of the women entering the labour force in the decade 1971–81, 60.6 per cent, were 15 to 34 years of age (Marshall 1987). Reports indicate that although women accounted for 70 per cent of total employment growth between 1976 and 1985, the proportion of all 25- to 34-year-old women in clerical occupations rose from 21 to 31 per cent, while the proportion of young women in service occupations remained constant at 21 per cent (Dumas 1986). One must also note the changing sectoral base of labour markets. The service-based industries – including trade, transportation, communications, utilities, insurance, real estate, business, community, and personal service – have grown much more rapidly than the goods-producing sector of the economy: they grew by 61 per cent between 1970 and 1985, while the goods-producing sector grew by only 13 per cent in the same period. Service industries commanded 66.5 per cent of all jobs in 1985, compared to 62.6 per cent in 1970 (Statistics Canada 1986 [autumn], 3).

Finally it is important to note the significance of part-time work in women's employment figures. Over one quarter (26 per cent) of all women in the paid labour force in 1991 worked part-time, up from 21 per cent in 1976. In comparison, just 9 per cent of employed men were working part-time in 1991. "In fact, women have consistently accounted for at least 70 per cent of all part-time employment in Canada over the past fifteen years" (Statistics Canada 1993 [spring], 3). Furthermore between 1975 and 1989 there was substantial growth in women's part-time employment in most occupational categories. The managerial and professional, clerical, service, transportation, materials-handling, and crafts occupations showed the most notable increases (Labour Canada 1991, 8).

In 1981 dollars part-time work averaged $6.84 per hour compared to an average of $8.64 for full-time jobs (Gee and Kimball 1987). A significant percentage of women working part-time – twelve per cent in 1976, and 27 per cent in 1986 – would have preferred full-time work, but it was unavailable to them. The high percentage of women who work part-time, and the difficulty women have in finding full-time work, is in some measure a result of society's expectation that women's primary responsibility is to home and family. Family obligations and demands often result in a lack of continuity in the labour force, with the result that women lose income, pension credits, and labour-market experience with no compensation in the form of formal recognition for work done in the home. The opportunity costs for women are also high. The Canadian Advisory Council on the Status of Women states that programs to alleviate this problem should be designed so that traditional stereotypes are not reinforced. The Council's 1987 publication on the Canadian Jobs Strategy notes, for example, that the majority of women in this federal program are still receiving training in traditionally female and lower-paying occupations (McKeen 1987).

ENTREPRENEURISM

This account of the labour-force participation of women, their wage and benefits situation, and the double workload of paid and unpaid work responsibilities that most carry helps to explain the burgeoning interest on the part of women in starting and maintaining their own businesses. These women are generally called entrepreneurs. Lavoie (1988) offers this definition of a woman entrepreneur: "a woman who has, alone or with one or more partners, started up, bought or inherited a business, is assuming the related financial, administrative, and social risks and responsibilities, and is participating in the firm's day-to-day management." Although this broad definition ignores notions of entrepreneurs as innovators, or indeed as employers, it is useful in this discussion since it relates well to our current knowledge about businesses owned and operated by women.

Lavoie's work recognizes the paucity of research on women entrepreneurs. She notes that, as of 1988, statistics in existing Canadian studies were not comparable or necessarily national in scope. With these caveats in mind, the following general conclusions about these working women is nevertheless interesting and perhaps indicative of the ways in which women try to maintain needed earnings and, at the same time, take more control of their working conditions.

According to Lavoie's findings, women entrepreneurs go into business for four major reasons: to work with their spouses; to make greater use of their talents than is possible in other employment; to be financially independent; and to take on a challenge. Over 60 per cent are married and most have a higher education level than both other women and most male entrepreneurs (a Saskatchewan-based study indicated, however, that male entrepreneurs in that province have higher education standing than their female counterparts). Businesses run by women tend to have a lower failure rate than those operated by men, but women entrepreneurs tend to earn lower incomes than men. This may be because their businesses are younger and smaller, because women invest less start-up capital, and because few women choose to enter the manufacturing sector, where higher incomes and profits can be made. The story behind these statistics strongly suggests that women have a more difficult time getting credit, fewer contacts in the business world, less experience than men, and less access to information and advice. They may therefore concentrate on forming smaller enterprises. Given these hurdles, the latest figures from *Canadian Social Trends* are not surprising: "Women are less likely than men to be self-employed. In 1991, approximately 525,000 women worked for themselves, representing just 9 per cent of all female employment. This compared with almost 1.3 million self-employed men, accounting for 19 per cent of total male employment. As a result, women represented only 29 per cent of all self-employed workers in 1991, a figure well below their share of total employment (45 per cent)" (Statistics Canada 1993 [spring], 5). Some recent articles and background papers suggest that women begin their own businesses out of frustration with the systemic barriers they meet in their regular jobs. They speak of the "glass ceiling" – the invisible barrier to recognition and promotion; the lack of flexibility – which inhibits women from balancing their family and paid work responsibilities; and the lack of autonomy: "We have termed the resulting isolation 'the glass box'. Women entrepreneurs are often isolated in a 'glass box'. They are surrounded by opportunities but lack the time, the resources, the know how, and the contacts needed to gain access to these opportunities" (Belcourt et al. 1991, i). Others are starting their own businesses in response to government initiatives in the privatization of social services, which employ a high percentage of women. According to Belcourt et al. (1991, i), "women are starting businesses at three times the rate of men." Most of the women in the study started businesses in the retail and service sector, most are married with children, and most work long hours to meet their business, household, and family responsibilities. And, as a result, most "are somewhat isolated from support networks such as friends and women's associations" (i). The report also found that women experience exclusionary and even discriminatory treatment in the business world. Women reported differential treatment from customers, suppliers, and employees.

SOCIAL PROGRESS TOWARD EQUALITY

It is evident even from this cursory look at women's place in the waged economy in this century that, despite the impetus of symbolic gestures such as IWD, women continue to struggle for appropriate recognition and compensation for their work, improved access to education, jobs and entrepreneurial activities, and much needed family supports. The notion that a woman's contribution to a family or a marriage is worth less than that of a man has been demonstrated by the lack of a pension or one of a smaller amount than the male's paid to her. Now, though, traditional views have been replaced by the concept of marriage as an economic and social partnership of equals. Yet, despite this dramatic shift in values, full social equality for women in families is far from ensured. All rights and responsibilities are not equally shared; nor are the consequences of particular acts or events such as the birth of a child, a separation, or a divorce. Crimes committed by one family member against another are not prosecuted and punished in the manner that comparable crimes against strangers would be prosecuted.

Social inequality becomes especially evident after the breakdown of a relationship. In general more women tend to become poor after a separation or divorce, while men tend to retain their previous economic status. Studies in the United States show that, after divorce, women's purchasing power is reduced by 42 per cent while men's increases by 70 per cent. The proportion of low-income families led by women in Canada has increased markedly over the past 25 years. In 1961 only 13 per cent of poor families were headed by women; this increased to 17 per cent in 1969 and to an estimated 37 per cent in 1985. Moreover true equality within families will not occur until the ongoing problem of wife battering is resolved. Despite higher prosecution rates for batterers and increased funds for transition homes, CACSW's 1987 study on wife battering shows that its effects on both women and children continues to be a pressing problem (McLeod 1987).

Furthermore no real progress toward social equality will be complete until the educational needs of women are recognized and accommodated. In the early 1890s the educated woman was a figure of fun and called a "bluestocking." Knowledge and matters of the intellect were considered essentially masculine concerns, and the woman who pursued an education was thought to have compromised her femininity. At one time, a high school diploma was all that was necessary to get a good job, and education beyond high school was largely a male preserve. Today, times have changed. At the beginning of the 1970s only 38 per cent of undergraduate degrees were awarded to women; by 1985, this figure had jumped to 52 per cent. We should also be encouraged by the changing profile of the enrolment of women in universities: between 1976 and 1984 the proportion of women enroled full-time in commerce and business management almost doubled, moving from 23.3 per cent to 42.3 per cent (Dumas 1986). Beyond the bachelor's degree, however, education is largely a nontraditional experience for women. In 1980, for example, only 37 per cent of masters' degrees and only 23 per cent of doctoral degrees were obtained by women. In 1985, 42 per cent of masters' degrees and 26 per cent of doctoral degrees were earned by women (Labour Canada 1987).

It is interesting to look at the relationship between education and labour-force participation. Men and women with less than grade 9 education have the lowest labour-force participation rates of any education category; women with less than grade 9 education have the lowest participation rates in Canada. Low educational attainment is also associated with low levels of income. The 1984 median income of families with a head having less than grade 9 education was $24,000, just 48 per cent of the median income of families headed by a university graduate (Statistics Canada 1987 [spring], 2–7).

The importance of education to waged employment is underlined when we consider the ways in which microelectronic and computer technologies are changing the labour market; indeed, a transformation of our economic base is occurring, along with the displacement of certain categories of work. The Conseil de la Science et de la Technologie (Québec) argues that computerization of service sector jobs results in staff reduction or increased levels of work without a parallel increase in jobs. This is especially true in banks, insurance companies, and supermarkets. Women are vulnerable in this labour-market situation, because traditional jobs are threatened and nontraditional jobs can present a major challenge, even though they offer higher pay and status. University enrolments show that women are responding to the challenge: fields that have been considered nontraditional, such as administration, applied science, and medical science programs, show increased female enrolment.

Canada's need for scientists appropriately prepared for current needs means that women as well as men will have to fill these jobs. Nevertheless women still face many hurdles within and outside of educational institutions. Cases brought before the Quebec Human Rights Commission by women point to systematic discrimination manifested in difficulties in getting hired, undervaluation of credentials and experience, use of nonpertinent criteria in the hiring process, inferior remuneration, orientation towards jobs with little advancement potential, isolation in a predominantly male environment, sexism, and sexual harassment. Ultimately we must look towards an acceptance and implementation of principles of equality. Women need equal opportunity to realize their goals without constraint of choice, and they deserve equal pay for work of equal value and the correction of past inequities.

CONCLUSION

The goals that motivated women on this continent and in Europe to militate for women's rights at the beginning of this century are the same goals that motivate us today: equal involvement of women in the political process; recognition of women's work in the home; fair treatment for women in the labour market; economic self-sufficiency and equality, dignity and fairness for all. Progress has been made since the first International Women's Day. There is now legislation in Canada on employment equity, employment standards, unemployment insurance, and human rights. Occupational opportunities are opening for women. There is no cause, however, for being smug; much remains to be achieved. Women still earn only 70 per cent of men's salaries (*Globe and Mail*, 24 December 1988, sec. A, p. 6) and are employed in a very restricted number of occupations. A significant number of women are forced to take part-time jobs, with lower wages and reduced benefits, when they would prefer full-time jobs. Maternity and parental leave provisions are still inadequate in many industries. Less than 10 per cent of children requiring day-care have access to licensed (which means supervised) care, and sharing of family and household tasks between spouses is still a goal rather than a reality. Whatever progress has been achieved by women has come as a result of their mobilization and their critical thought – women have challenged the status quo, articulating and giving names to issues (suffrage, maternity leave, pay equity), and women academics have provided valuable theoretical frameworks. For our next leap forward we need to build on this critical thought of women. The received wisdom of the past, particularly as it related to gender roles and women's place, was developed and articulated by men based on male experience; however, as we learn together about the real lives of women and men, we become part of the movement for change. Just as International Women's Day was inspired by social conscience and a search for justice, so our lives, too, must be motivated by these important goals.

NOTES

I wish to thank Katie Pickles for her assistance in gathering data for this paper.

REFERENCES

Canada. Employment and Immigration Canada. 1989. Benefit entitlement directorate (July). Unpublished data.

Canada. Labour Canada. 1987, 1991. *Women in the labour force.* Catalogues L38-30 and LO16-1728/90E. Ottawa.

Canada. Statistics Canada. 1986, 1987, 1989, 1990, 1992, 1993. *Canadian Social Trends.* Ottawa.

– 1988a. *Historical labour force statistics.* Catalogue 71-201. Ottawa.

– 1988b. *Earnings of men and women*. Catalogue 13–217. Ottawa.

Canadian Advisory Council on the Status of Women (CACSW). 1987. Progress toward equality for women in Canada. Notes for a presentation before the Standing Committee on the Secretary of State. Ottawa.

– 1987. *Integration and participation: women's work in the home and in the labour force*. Ottawa.

Belcourt, Monica, Ronald J. Burke, and Hélène Lee-Gosselin. 1990. *The Glass Box: Women Business Owners in Canada*. Ottawa: CACSW.

Canadian Congress for Learning Opportunities for Women (CCLOW). 1986. *Decade of Promise: An Assessment of Canadian Women's Status in Education, Training and Employment, 1976–85*. Toronto: CCLOW.

Conseil de la science et de la technologie. 1986. *La participation des femmes en science et technologie au Québec*, document no. 86.06. Québec.

Côté, Renée. 1986. *La journée internationale des femmes*. Montréal: Les Éditions du remue-ménage.

Day, Tanis. 1987. *Pay equity: some issues in the debate*. Ottawa: CACSW.

Dumas, Cécile. 1986. "Occupational trends among women in Canada: 1976 to 1986." In Statistics Canada, *The labour force*, 85–123. Ottawa.

Gee, Ellen M. and Meredith Kimball. 1987. *Women and Aging*. Toronto: Butterworths.

Goldman, Emma. 1969. Rpt. "Women suffrage." Ch. in *Anarchism and Other Essays*. New York: Dover Press. Orig. pub., 1917.

Hawke, Kelly. 1988. "Political women: their ranks have grown, but some say not by enough." *The Financial Post*, 12 December, Spectrum 2, p. 13.

Kesterton, Michael. 1988. "Ringside at a cat fight: why women can't win." *The Globe and Mail*, 4 January, p. 84.

Lavoie, Dina. 1988. *Women Entrepreneurs: Building a Stronger Canadian Economy*. Ottawa: CACSW.

Marshall, Katherine. 1987. *Who are the professional women?* Report prepared for Statistics Canada (catalogue 99-951). Ottawa.

McKeen, Wendy. 1987. *The Canadian Jobs Strategy: Current Issues for Women*. Ottawa: CACSW.

McLeod, Linda. 1987. *Battered But Not Beaten: Preventing Wife Abuse in Canada*. Ottawa: CACSW.

6 "No Skill Beyond Manual Dexterity Involved": Gender and the Construction of Skill in the East London Clothing Industry

ALISON KAYE

The title quotation (Stedman Jones 1984) refers to the nature of women's employment in the East London clothing industry of the late 1800s, work thought to require little skill or ability other than manual dexterity. In many respects the industry (which here includes the clothing and suede and leather industries) appears to have changed little since the last century. Today female employees occupy a position similar to unskilled machinists: for example as homeworkers, they sew together coat linings and earn as little as 10 pence per lining. The structure of the industry, a plethora of small firms and outwork units that rely upon a backbone of homeworkers, is still characterized by labour-intensive methods of production, cheap migrant[1] labour, and seasonal, casual labour patterns.

Like the Jews who fled Eastern Europe at the turn of the century, Bangladeshis,[2] who began arriving in Britain during the 1960s, are the most recent migrants to have taken up employment in the industry. They live and work in proximity to the heart of the trade in the area of Spitalfields, which incorporates Whitechapel and Aldgate in what is now the East London borough of Tower Hamlets. Despite their low pay, appalling working conditions, and entry into the trade at the lowest level as unskilled or semiskilled workers, this group has found the garment industry an important source of employment. Bangladeshi men have taken jobs as machinists, a task traditionally considered to be female and unskilled yet often abandoned by White women because of low pay. Many Bangladeshi women also do unskilled machine work, but, unlike the men employed in outwork groups or factories, they work at home alone; they also earn less pay. As Morokvasic (1987) notes in her study of migrant workers in the contemporary Parisian garment industry, males are much more likely to advance in the industry and go on to skilled tasks as well as to find relatively stable employment in the "official" sector of the industry.

Women in general are discriminated against in the British labour market (Beechey and Whitelegg 1986), but migrant women are also subject to racism and often occupy the worst-paid, least-secure jobs, as a recent article by Breugel (1989) clearly illustrates. Recent studies have begun to redress the lack of empirical and theoretical knowledge of migrant women in the labour force. Westwood's (1984) workplace study of women working in a multiracial environment in a hosiery factory is one example. Other recent studies of the working lives of specific groups of women include the study of Greek-Cypriot women in the north London clothing industry by Anthias (1983); the work of Phizacklea (1982) on Afro-Caribbean women; and, more recently, Baxter and Raw's work (1988) on Chinese women in the catering industry. All of these studies illustrate the racism and difficulties faced by migrant women in the labour force and show that working-class migrant women are limited to the lowest-paid, least-secure, low-grade manufacturing and service jobs.

Women have always occupied semiskilled and unskilled positions, while men have traditionally been employed as skilled workers. Over the years the definition of skilled work has changed, but the gendered division of skill has persisted. In the course of this essay I will examine these changes and the links between gender and race in bringing about some of these alterations. An historically sensitive approach is adopted in the first of the sections to allow illumination of present-day definitions of skill in the local industry, which consistently has been redefined and has allocated roles according to gender and racial divisions in the labour force. The importance of gender divisions within the working class throughout most of the last century illustrates how material practices and ideology are inextricably linked and that the continual perception and definition of women as unskilled workers is strongly linked to gender ideologies and to notions of masculinity and femininity that are firmly embedded in the production process. This relationship becomes more complex when migrant men take on "women's work" and thus are instrumental in bringing about shifts in skill categories. In the second section I will draw upon interviews carried out with homeworkers and others involved in the clothing industry, where Bangladeshi women are now the least-skilled, worst-paid workers.

Finally I question critically the social definitions of skill and look at how those categories may be in the process of changing once again.

CONTESTED TERRAIN:
GENDER AND SKILL

The history of the East End "rag trade" has been written many times (e.g., Hall 1962; Stedman Jones 1984; Schmiechen 1984) and so will not be repeated in detail here. An outline of the structure as it developed after the 1830s, however, is necessary in order to illustrate the characteristics of the local industry. During the course of the last century, the industry underwent tumultuous changes in its structure, organization of production, and composition of the labour force. The impact of industrial capitalism in its move towards a consumption-based economy included a rise in the production and demand for ready-made clothes, which changed the industry by replacing craft-based production with labour-intensive methods. In other words the skilled artisan base began to be eroded, just as the one-time "make through" master tailor craft was increasingly divided into many different tasks, resulting in the division of labour and rise in the incidence of deskilling.

As early as the 1820s the trade was threatened by work being given out to groups of outworkers and women (Taylor 1983), the cheap labour pool that would include migrant Jews later in the century. This method of production gave rise to the development of "slop" clothing manufacture (cheap and ready-made) and gave rise to the terrible sweated conditions of labour, the "low wages, long hours and insanitary working conditions" (Morris 1986) endured by inexpensive female, child, and migrant Jewish labour. The organization of the industry was spatially dispersed and disorganized; there was manufacture without factories. A maze of small outwork shops dominated Whitechapel and Aldgate, and more people were employed as home-workers in the many nearby run-down residences. Much of this industry relied upon subcontracting, which was seen as a scourge by many workers who argued that it increased further the division of labour (Schmiechen 1984, 190). A yet more contentious issue that divided workers was the introduction of waged female labour, which was seen as a threat to the safety of male jobs and the power of the all-male preserve of the tailors' union.

The London Operative Tailors was one of the strongest unions in the country prior to 1830, and its members were able to control the supply of labour as well as to negotiate wages successfully (Schmiechen 1984, 8). According to Taylor (1983, 101), "the tailoring workshop of the eighteenth century had been a man's world of hard work, hard drinking, and tough union politics." This "man's world" was inhabited by "flints," some of the best-paid and most highly skilled workers in the country, who worked in the "honourable" (nonsweated) sector of the trade. Prior to the rise of slop manufacture they were able to keep the growth of the less honourable sector, that producing piecework and consisting of homeworkers, at bay. Even the language then used to describe the different sectors is indicative of the status and respect enjoyed by artisans and the disdain with which other, nonskilled workers were viewed. Women were brought into the trade on a waged basis, though only to perform subordinate tasks such as buttonholing or sewing on buttons – jobs that neither the tailor nor his apprentice would perform (Morris 1986). Girls came to the trade at an early age but never on the same terms as their male peers; both performed menial tasks such as running errands, but girls were not allowed to learn the same skills as boys, who would go on to become master tailors. Knowledge was passed on from the male adult to the young male via privileged access, and the latter would "absorb the mysteries of the craft" (McClelland 1989). Restricted access to apprenticeships was a common form of maintaining privileged skilled status (Rose 1986) and controlling who did and did not enter the trade.

The technological developments of the sewing machine and the bandsaw – machines that allowed huge increases in production – played an important part in the process of deskilling. The sewing machine, for example, made it possible to complete between one to two thousand stitches per minute on a shirt, a considerable advance on the thirty-five permissible by hand (Schmiechen 1984, 25), thus considerably increasing and intensifying production. The machine was designed and made with women in mind and, as this excerpt from an early advertisement illustrates, it was celebrated not only as a labour-saving instrument but as one particularly suited to female usage: "We must not forget to call attention to the fact that this instrument is peculiarly calculated for female operatives. They should never allow its use to be monopolized by men" (Cooper; quoted in Chenut 1983, 68). The design was equally suited to home and factory use. In the East End it was used largely as a domestic machine, and women bought or, more often, rented machines so that they could work at home. Many households in the area possessed a Singer sewing machine by the late 1880s: "The whole of East London by 1888 was mapped out in sales districts with regular armies of collectors ... to collect the installment payment ... The Singer Company had thirty collectors in the East End alone" (Schmiechen 1984, 26). The machine was quickly adapted for home production and women became engaged as waged workers within the so-called private sphere. At the time it was thought that the industry, based as it was on small workshops and domestic production, was backward and that sweating would eventually disappear. Marx, writing in the 1860s, stated that the sewing machine would cause the industry to develop into the "factory system proper," wherein all the tasks would be concentrated under one roof and management (1976, 603). Yet Marx failed to recognize the import of a cheap, unskilled female labour force and the fact that, if employers could continue to enjoy profits by using labour-intensive methods of production, then there would not necessarily be a move to centralize production. Indeed, the present-day industry remains spatially dispersed, with some factory production but also with many small outworking units and homeworkers.

The development of the East End industry highlights the unevenness of capitalist development and throws into question the traditional division between home and work. Clearly this industrial pattern does not fit the model of the public world of work and the private sphere of the home but points instead to the variety of changes in the production process, which, in this instance, maintained a strong link between domestic and factory production. While women were drawn into waged labour the family continued to exercise an influence upon the productive activities of its members (Tilly and Scott 1987, 232), and women who were seen primarily as secondary wage earners were restricted to particular jobs.

The entry of cheap female labour into the slop trade threatened the status of skilled male workers. Traditionally the tendency toward deskilling and the introduction of unskilled female labour have been seen as weakening forces affecting labour's bargaining power with capital: "this excessive addition of women and children to the working personnel," writes Marx, "at last breaks the resistance with which the male workers had continued to oppose the despotism of capital throughout the period of manufacture" (1976). While it is undoubtedly correct that female labour weakened the resistance of the tailors' union and that employers used it to their own advantage, the situation brought to the surface the already underlying tensions and contradictions between male and female members of the working class; as discussed above there was already a strongly demarcated division of labour in an industry that actively prevented women from gaining recognized skills and that restricted them to only the most menial, unskilled tasks. This effectively meant that women were restricted to subsistence wages, and were forced to work for piece-rates, often for long hours, as homeworkers. Due to the union's effectiveness in ensuring its members' rights, male trade union members worked fewer hours than women despite legislation restricting the number of hours women should work (Morris 1986, 78). The question of female labour was a contentious issue in the union and caused a division amongst tailors (Taylor 1983, 115); women's labour was seen to be undercutting male jobs. Mayhew (1985) recorded tailors' reactions towards the influx of women into the industry, one of whom complained: "Formerly an operative tailor's wife never helped him ... [H]is wife attended to her domestic duties. The decline in the prices of our trade arises, in our opinion, from our wives and daughters being brought to work, and so to compete with ourselves." Not only were women in competition with men, but the order of the gendered division of labour was being threatened; women were perceived to be moving away from their domestic duties into waged work and flouting the dominant image of what it was to be a woman. The tailors were fighting not only to maintain jobs and privileged skilled status but also to hang onto the sexual division of labour and, ultimately, to maintain their gendered identity as men. If women (and, later, Jewish migrants) could not be prevented completely from entering into tailoring, then tailors could and would struggle to maintain their skilled status.

The tailors' actions – which included strikes against women workers because of the threat of cheap female labour and the introduction of piecework (Taylor 1983, 102); strikes to establish independence from employers; and denouncement of Jewish fellow workers who were said to be undercutting costs (Schmeichem 1984, 35) – illustrate some of the complexities of gendered identity and the many overlapping layers that contributed to both the external and internal meaning of what it was to be a White, English, Christian male. The objections raised to women entering the trade reveal the tensions and divisions that existed within the working class and the differential status of men and women within it. The migration of Jews later in the century was also seen to be a threat to the economic and social position of skilled, White, English males. The tailors' protests become a struggle centred around the perceived challenge to their racial, class, and gender identities, each of which was equally important: "differences between white and Black, middle and working class, Christian and Jew are no less differences than the one between girl and boy" (Spelman 1990, 97). Gender was not the sole criterion in the tailors' efforts to maintain their skilled status; they clearly saw themselves as superior to the Jewish male, too, for part of their gendered identity and self-esteem rested upon their conception of themselves as White, Christian males.

In her study of the nineteenth-century printing industry, an all male preserve, Cockburn (1983) analyzes the printers' strong identification with work as an important aspect of their self-identification as men.

Similarly, as craftsmen, tailors invested a great deal in their work, which was a means of earning a livelihood and a source of good pay. It was also an important part of their self-image: from boyhood through to the completion of their apprenticeship, tailors identified themselves as skilled workers, as artisans who, moreover, were able to support their families without their wives' needing to take up employment. They were honourable workers. The strength of their identification as men came in part from their membership in the union, which, as McClelland (1989) shows, allowed them to be visible and to assert both independence and family dependency. It was partly this ability of the unions to shape self-image that made them the class institutions they were. And all of this was threatened by the influx of cheap labour.

At the end of the 1800s and into the beginning of this century, Jewish migrants moved into the industry. Although many of the men brought tailoring skills with them they were designated as semi- and unskilled workers because, despite their skills, they had little control over the employment they were forced into accepting. The majority worked as outworkers, supplying their own machines and relying upon subcontracting as a means of getting work. The skilled tailors saw the new migrants in a way similar to that in which they viewed female labour: as a threat to the ever increasing subdivision of labour and as a cause of the sweated conditions that were rife by the 1880s in the East End rag trade. Despite the depressed state of the industry into which migrants flowed, they were nevertheless held responsible for the rise in casual, unskilled, cheap labour (Stedman Jones 1984, 110), though not by all sections of workers (Schmiechen 1984, 105). In many instances the pay and working conditions were so abysmal that the only ones willing to accept work were those who had no alternative, namely Jews and women. Jewish males, faced with the hostility of indigenous male workers and the immediate need to feed and sustain their families, were forced into accepting conditions that "placed their masculinity on a precarious foundation" (Baron 1989, 190). Job opportunities were confined to the unskilled and least-permanent tasks, typically machining work, which was defined in the industry as women's work.

The struggle of Jewish males to have their work recognized as skilled paralleled that of the tailors'. Both were reassertions of masculinity in the face of racial and gendered degradation. The dual effects of subordination in the workplace and the concomitant struggle to maintain patriarchal status within the family and community led to attempts by Jewish males to become autonomous and to be considered skilled within the workplace. This was partly achieved throughout the East End industry by groups of males who set themselves up in outwork units to take contracts on a "cut, make and trim" footing from larger manufacturers (Kosmin 1979; Phizacklea 1990, 26–8). As Kosmin (1979) points out, this was quite easy to accomplish in the spatially diverse, labour-intensive London industry, with its lack of large-scale factory production. Organization into outwork units enabled Jewish men a certain autonomy over the labour process, much as organizing into unions such as the Master Ladies' Tailors Organization (Phizacklea 1990, 26–8) did for women.

Birnbaum, as Phillips and Taylor (1980) note, outlines another mechanism by which men asserted their masculinity at work. In 1926 the Wages Council, responsible for setting wages and standards in the clothing industry, set about defining the limits of skill. Despite the influx of women into the industry in the twenties (Phizacklea 1990, 27), or, rather, because of it, men managed to lobby for their work to be defined as skilled while that of female machinists remained classified as semiskilled. Birnbaum, according to Phillips and Taylor, maintains that the actions of the men were rooted in the need to assert their masculinity within the family, for "craft status was identified with manhood" (Phillips and Taylor 1980, 85) and closely associated with their status as family breadwinners. The struggle in the workplace was intimately bound up with their place in the patriarchal Jewish family, the status of which was maintained and enhanced by that of being a skilled worker.

The need and struggle of Jewish men in the local industry to establish themselves as skilled workers cannot be separated from the gendered ideology of work, either. To be a skilled worker meant not only status and better pay but affirmed manhood and the maintenance of patriarchal power both at the workplace and in the household.

SHIFTING GROUND?
CONTEMPORARY SKILL DIVISIONS

Bangladeshis migrating to London from the 1960s onwards have found themselves in the East End, living almost street for street where former migrants once lived; likewise they are often reliant upon the rag trade as an important, if not the only, source of income. Despite a 61 per cent job loss in the clothing industry of Tower Hamlets during 1971–81, over 33 per cent of Bangladeshis registering as unemployed in 1983 stated that their last job had been in the clothing industry (Buxton 1986). This figure excludes many Bangladeshi women who never register as unemployed; they are particularly vulnerable, because many are homeworkers employed in insecure, ill-paid, unskilled jobs as machinists (Mitter 1986). The structure of the industry has changed little since the turn of the century, as noted earlier; it is still based on small workshop and domestic production and relies upon labour-intensive methods of production. Employment remains seasonal, casual, low paid, and irregular. While production in other areas of the country is turning over to computer aided design and "hi-tec" (Rainie 1984), the East End trade is indicative of the uneven transformation of the capitalist production process.

The industry is a complex structure of subcontracting firms that operate on a CMT (cut, make, and trim) basis, receiving orders from large retailers and local suede and leather shops. CMT firms receive the design and material and then begin to put the required garment together. This is a highly divided process and one garment may be contracted out, perhaps six or seven times, to various firms so that workers never see the final product. Few Bangladeshis own CMT firms; the majority are employees. Males work in a group of six to eight and compete with other outwork groups for contracts. At the very bottom of this ladder are the homeworkers, the majority of whom are women.

The eight women who took part in my preliminary study originally came from rural areas of the state of Sylhet, like the majority of Bangladeshis living in the East End.[3] And, like the majority (Werbner 1987), all of the women were born in small villages, mainly into middle-income, rice-farming families. All but two of the women had spent their entire lives prior to migrating in rural areas; one had attended school in Sylhet Town, the most important urban area in the state, while the other had lived with her brother and sister for a while in the town. Six had been educated to the primary school standard and then been withdrawn from school at age twelve, while two had had further schooling. The majority of girls in Bangladesh, especially in rural areas (Jahan 1982), do not go to school beyond the primary level for a variety of reasons. Some enter *purdah* at puberty, which restricts their physical and social movements until they are married at fourteen or fifteen. Financial restrictions often mean that education cannot be continued, because secondary education, unlike primary education, is not free.[4]

Most of the women had been married two or three years after completing their schooling. At the time of interviewing all of the women were married, though they had not necessarily been married while they were homeworkers. The women came to England to join husbands and fathers; two came with their mothers, and the others lived in the homes of their in-laws, having moved there upon marriage. Although the women fit quite closely into the standard picture of homeworkers as married women with young children, these conditions were not the only criteria for working at home. One, a divorcée whose only child was in Bangladesh, worked to support herself while another had worked as a schoolgirl during the holidays to earn pocket-money. The six remaining women were married with between three and seven children, who varied widely in age from under five to over twenty years. For all of the women homeworking was their first experience of paid work, though not necessarily their first experience of work. Two of the women had carried out arduous agricultural processing work as well as heavy-duty household tasks. The rest, who were younger, had not needed to carry out such heavy agricultural work due to the mechanization of certain tasks. Some had had servants in Bangladesh to carry out the more unpleasant household tasks. The economic pressure on these women to work due to husbands' low wages and unemployment was an initial shock, and the added burden of child care on top of their paid work was, most women thought, something that would never have happened in Bangladesh. Homeworking was not something these women did just so that they could stay at home with their children; since opportunities open to them were limited, there was little choice left but homework.

Having rented or bought an industrial sewing machine, often paid for by funds from the entire family, such women work in isolation at home as machinists. The industry needs to respond quickly to fashion, which is seasonal and increasingly changeable, and the employers take advantage of outworker and homeworker flexibility in that they can be disposed of when there is no work, unlike full-time factory staff. Moreover they are an immensely inexpensive source of labour. From my sample it is possible to estimate what an average homeworker's weekly wage might be. Parveen, for example, earns the going rate for coat linings without pockets, 10 pence per lining. Her wage varies according to the supply of work and the amount of work she receives; it is possible that in any one week she can sew between 200 and 400 linings, her wages thus varying between £20 and £40 per week.

Such low-paid work amongst migrant women is not an isolated occurrence. At a recent European Community seminar it was stated that migrant women throughout Europe were characteristically employed in low-paid, unskilled jobs with little security and in poor working conditions (de Troy 1987). Minority women are more likely than native-born White women to be employed in the low-paid declining manufacturing sector of the Western European economies. This is evident in the clothing industries of Paris, France (Morokvasic 1987), and of Britain and West Germany (Phizacklea 1987). Within the British industry Black women are overrepresented and comprise over 80 per cent of the work force (Coyle 1982), mainly in semi- and unskilled jobs. Migrant women enter a labour force demarcated by gender and stratified by race where they hold low-paid, low-status jobs and work extra hours for less pay (Breugel 1989). Bangladeshi women and men occupy unskilled and semiskilled jobs, but women are more likely to do the former.

Yet just how unskilled is the work performed by homeworkers? In the East End trade, sewing coat linings is considered to be one of the least-skilled of jobs.[5] That this is the case is proven by the low piece-rates of 10 pence per lining, 25 pence with pockets. Constructing a lining involves sewing together the front, back, sleeves, and pockets, which varies in difficulty depending upon the number and cut of pieces and the type of material. Shiny lining material, for instance, is more difficult to control because of its slipperyness. Defining this work as unskilled, therefore, does not mean that it requires no skill. As an officer in a local clothing organization noted, no job in clothing is unskilled; all tasks require skill. But unless a task is officially recognized as skilled (Lever 1989), the worker has no power or status to demand more pay or better working conditions. Futhermore the categorization of tasks as skilled or unskilled does not necessarily have any bearing upon the type of work carried out but is likely to be associated with the gender of the worker.

Employers offer no formal training to new homeworkers, nor do they necessarily require an employee to possess sewing skills. Indeed, of the homeworkers in my preliminary study of eight, only one had any previous knowledge of sewing gained either through making the traditional *kantha*[6] or by attending local training courses in textiles. The majority of women knew neither how to sew a lining nor how to operate an industrial sewing machine. All of these skills were acquired on an informal basis, usually via the relative, friend, or neighbour who had found them the work; Fatima's brother, for example, a machinist in a factory, taught her how to use the machine. This pattern is repeated again and again: women rely upon their informal kinship and friendship networks both to find employment and to act as a training source. Employers recognize this and rely upon the women's contacts to carry out training. Hafiza told me that she recruited and trained her neighbours for homeworking at no extra cost to her employer. Homeworkers do not always have such direct contact and are sometimes left to their own training devices; in such instances women are left a sample of the item to be sewn with their first delivery of work and are expected to learn from the sample: "I was left a sample to follow, I could ring the employer if I wasn't sure how to do it and ask for advice on the phone," said one woman. But this can often be time consuming and difficult: "at first I had to keep redoing them until they were right ... [T]he first time is hard, you don't know anything." First-time homeworkers all had the same story to tell – of hours spent at the machine sewing, unpicking and redoing linings, slowly learning the skill of successfully operating the sewing machine, and putting together linings with both speed and accuracy. Throughout the East End industry, both inside and outside the factory, a period of informal training at little or no cost (in the case of homeworking) is a recognized feature of becoming a machinist. The rationale for allowing such informal training is usually that workers are not reliable and that if an employer were to waste time and money on a formal training period the worker would then move to a higher-paid job with the newly acquired, official skill.[7] Not only does this practice keep costs and wages low and prevent workers from gaining recognized skills to give them greater power in the labour market, but it continues to perpetuate myths and ideologies surrounding women's work.

Contemporary feminists writing on the issue have argued that it is the sex of the worker rather than the content of the job that determines the definition of the task as skilled or unskilled (Philips and Taylor 1980). Locally specific studies of homeworkers such as Mies's of lacemakers in India (1982) and Lever's of embroiderers in rural Spain (1988) have focused upon the importance of the ideology of the housewife and the consequent invisibility of women's skills outside of the home as contributing factors to the problem. While the studies cited are historically and locally specific, Elson and Pearson (1984) have provided a more general materialist interpretation of the social relations that contribute to the inferior position of women at work. In their analysis of "world market factories" and the predominant employment of women as unskilled workers in developing countries therein, they ask why it is mainly women who are so employed. Part of their answer, after interviewing companies that employ such women, is that women are predetermined as "inferior bearers of labour" as a result of the training they have received at home in preparation for being a woman. Interestingly for the discussion here, Elson and Pearson focus upon the feminization of sewing and the sewing skills learned at home as being easily transferable to paid employment involving the same or similar tasks: "since industrial sewing of clothing closely resembles sewing with a domestic sewing machine, girls who have learnt such sewing at home already have the manual dexterity and capacity for spatial assessment required" (Elson and Pearson 1984, 23). Although not all women possess these skills it is usually assumed that women naturally possess skills of this kind – even if, as in the case of the Bangladeshi women in this study, none of them were able to sew the *kantha* but typically relied upon a single female member of the extended family to carry out the making of the traditional quilt.

Sewing in particular is culturally identified as a feminine task and, as already discussed, the sewing machine is as much a domestic as it is an industrial object.

The possession of a skill, then, is not actually the issue when it comes to employment options for women; rather, it is the ideological construction of certain tasks as male or female that continues to control the East End clothing industry in terms of material practices. The influx of the two migrant groups has meant that at certain historical periods there has been a surplus of a cheap and ready labour. Homeworking has continued in part due to the continued perception of women as housewives and mothers, indeed as "bearers of inferior labour," especially when they are not White. Skill boundaries are not static; in the East End industry, skill boundaries and the gender of the person carrying out certain tasks are by no means a foregone conclusion. The Bangladeshi population has been faced with a lack of opportunities in the labour market due to the effects of racism. For men this is clearly demonstrated by the jobs available to them. Their apparent willingness to take on machining jobs, which is considered as women's work, was commented upon in amazement by White males currently involved in the economic development of industry (unpublished interviews). Towards the end of the last century, Jewish males constrained by racism and sexism had no choice other than to take up work that was perceived as feminine and that no White, Christian male would do; a similar process is now taking place in East London.

Today Bangladeshi males take on work that White women are no longer willing to do because of the low pay and generally poor working conditions. That migrant males have no option in the employment they are forced to accept is illustrative of their devalued status in British society; that society then conceptualizes their masculinity based on the form of employment open to them: women's work. In a generally patriarchal society there is no one form of masculinity, but several, just as there is no single manifestation of femininty. Patriarchal power bases its existence not only upon the subordination of women but also on the domination of White heterosexual males over others. This power is exercised in various aspects of society, including employment. Masculine work is considered as that which demands officially defined skills, dominance of machinery, and role fulfilment as the family breadwinner, a position that importantly enables the assertion of masculinity in the public sphere as well as at home. To be able to assert masculinity in this way is to reject subservience and embrace dominance. Bangladeshi males are not now in a position that enables them to assert their masculinity in this way, because the work they do is considered to be feminine; but it is likely that, like the Jews before them, this group will struggle to establish masculine status in the world of work, at least. Yet gender differentiation in the production process means that Bangladeshi men frequently have better chances of achieving skilled status and even of owning a small business than do Bangladeshi women.[8]

GENDER AND SKILL: HIERARCHIES IN THE EAST LONDON INDUSTRY

To return now to my title, and to Stedman Jones's original use of the phrase "no skill beyond manual dexterity involved," I have argued that manual dexterity is itself a skill. The association between femininity and manual dexterity has led to the latter being idealized as a lowly, female trait, one that is inherent in feminine nature; but, contrary to this ideological construction of women and the work they do, manual dexterity is learned, often painfully and at economic cost, as in the case of Bangladeshi homeworkers. Moreover there are inherent contradictions in the feminization of such skills: they are often performed by men, as well, such as those Jewish and Bangladeshi males who were and are expected to carry out unskilled machining tasks.

In relation to the local example pursued here, skill has consistently been defined in relation to gender and race, and the least-skilled tasks in the industry have been assigned to the groups with the lowest social status at historically specific moments: women, Jews, and Bangladeshis. At the most general level this tendency can be analyzed in a patriarchal framework, but such an analysis would not explain the complexity of the ideological and material practices at work that make skill divisions in the East London clothing industry different from elsewhere. I have found it analytically more useful to look at the social construction of skill by focusing upon gender and social relations together. Skill divisions, I found, are based upon and developed according to gender and racial ideologies and practices in which women and ethnic minorities have historically and contemporarily occupied the least-skilled positions. This categorization is a result of gender and racial hierarchies between men and women *and* between men and men. Those in relative positions of power have managed through protest and practice to maintain the definition of their work, or to redefine it, in the skilled category.

Women in the East London clothing industry have remained disadvantaged, however, in comparison to men. Bangladeshi women remain as homeworkers while males have easier access to permanent factory positions. The industry has continued to rely upon women's alloted role as housewives and upon their invisible skills, which lead them to unskilled, low-paid jobs to the benefit of the employer. Ideological and social factors have consistently contributed to this construction of women as unskilled workers. That this appears to have changed relatively little since the turn of the century requires that even more question be asked concerning the position of women in contemporary society.

NOTES

This paper was originally presented at a "Feminism and Historical Geography" conference held at University College, London, England, in November 1989. I would like to thank participants at that conference and Audrey Kobayashi for their helpful comments.

1 The term "migrant" is used to refer to people born outside of Britain who have moved to areas within Britain. In relation to Bangladeshis it is used specifically to refer to first generation migrants to emphasize the qualitatively different experience of their lives before and since immigration.

2 Bangladeshi is used interchangeably with Bengali; most prefer to call themselves Bengali or, occasionally, Sylheti. I have chosen to use Bangladeshi to avoid confusion with Bengalis from West Bengal and also to follow the pattern of recent literature (e.g., Carey and Shukur [1985], amongst others).

3 The majority of Bangladeshis living in Britain originate from Sylhet in the northeast of Bangladesh. In his paper "Salience of homeland," Fazlul Alam (1988) distinguishes two major groups of Bangladeshis in Britain: those who are part of the movement from Sylhet, and the educated, largely urban elite from Dhaka and from the group known as "mofussal graduates."

4 One woman in the study told me her father sponsored her friend through secondary education since her own family was too poor to do so.

5 Personal communication from a clothing trainer employed to teach courses run by the Inner London Education Authority.

6 A *kantha* is a traditional Bengal quilt, "a result and expression of the country's geographic and climactic character. The lightness and airiness of muslin counteracts the heat and humidity of ... the summer months ... women make *kanthas* of several layers of fine disused cloth, quilted together with embroidery" (Stockley 1988).

7 This was communicated to me in interviews several times by employers who decried the lack of training and the skill shortage in the industry.

8 Lever (1988, 12) quotes an interesting example from Spain. After the Civil War male unemployment was very high and many men had to turn

to new forms of employment. Within the embroidery industry, which Lever writes about, men took up ironing in the laundries, a job previously performed by women and classified as unskilled. Recently there has arisen a division of skill amongst male ironers; ironing carried out by hand is now considered to be more skilled than ironing done with machinery. Lever argues (and I would agree) that this is a result of men attempting to regain their masculinity by upgrading what was once women's work to the skilled, masculine status of men's work.

REFERENCES

Acker, J. 1989. "The problem with patriarchy." *Sociology* 23. 2: 235–41.

Alam, Fazlul. 1988. Salience of homeland: societal polarization within the Bangladeshi population in Britain. Research papers in Ethnic Relations, no. 7. Warwick: University of Warwick.

Alexander, S. 1976. "Women's work in nineteenth-century London: a study of the years 1820–1850". In *The Rights and Wrongs of Women*, J. Mitchell and A. Oakley, eds, 59–111. London: Penguin.

Anthias, F. 1983. "Sexual divisions and ethnic adaptation: the case of Greek-Cypriot women." In *One Way Ticket: Migration and Female Labour*, A. Phizacklea, ed, 73–94. London: Routledge.

Baron, A. 1989. "Questions of gender: deskilling and demasculization in the U.S. printing industry, 1830–1915." *Gender and History* 1. 2: 178–99.

Barret, M. 1988. *Women's Oppression Today*. London: Verso.

Baxter, S. and G. Raw. 1988. "Fast food, fettered work: Chinese women in the ethnic catering industry." In *Enterprising Women: Ethnicity, Economy and Gender Relations*, S. Westwood and P. Bhachu, eds, 58–75. London: Routledge.

Beechey, V. and E. Whitelegg. 1986. *Women in Britain Today*. London: Macmillan.

Berg, M. 1988. "Women"s work, mechanization and the early phases of industrialization in England." In *On Work: Historical, Comparative and Theoretical Approaches*, R.E. Pahl, ed, 61–94. Oxford: Basil Blackwell.

Birnbaum, B. n.d. Women, skill and automation: a study of women's employment in the clothing industry, 1946–1972. Unpublished paper.

Black, C. 1907. *Sweated Industry and the Minimum Wage*. London: Duckworth and Co.

Black, C. 1983. *Married Women's Work*. London: Virago.

Blumenfeld, S. 1988. *Phineas Khan: Portrait of an Immigrant*. London: Lawrence and Wishart.

Bruegel, I. 1989. "Sex and race in the labour market." *Feminist Review* 32: 49–68.

Buxton, T. 1986. *Off Our Back*. London: Aldgate Press.

Carey, S. and A. Shukur. 1985. "A profile of the Bangladeshi community in East London." *New Community* 12.3: 405–17.

Chenut, H. 1983. "The sewing machine." In *Of Common Cloth: Women in the Global Textile Industry,* W. Chapkis and C. Enloe, eds, 68–70. Amsterdam: Transnational Institute.

Cockburn, C. 1983. *Brothers: Male Dominance and Technological Change.* London: Pluto Press.

Connell, R.W. 1987. *Gender and Power.* Cambridge: Polity Press.

Coyle, A. 1982. "Sex and skill in the organisation of the clothing industry." In *Work, Women and the Labour Market,* J. West, ed, 10–26. London: Routledge.

De Troy, C. 1987. Migrant women and employment. Community seminar, final report, 17–18 September, Brussels.

Elson, D. and R. Pearson. 1984. "The subordination of women and the internationalisation of factory production." In *Of Marriage and the Market: Women's Subordination Internationally and its Lessons,* K. Young et al., eds, 18–40. London: Routledge.

Gannagé, C. 1986. *Double Day: Double Bind.* Toronto: Women's Press.

Hall, P.G. 1960. "The location of the clothing trades in London, 1861–1951." *Transactions and Papers of the Institute of British Geographers* 287.

Jahan, R., ed. 1982. *Women in Asia.* London: Minority Rights Group.

Johansson, E. 1989. "Beautiful men, fine women and good workpeople: gender and skill in northern Sweden, 1850–1950." *Gender and History* 1. 2: 200–12.

Kosmin, B. 1979. "Exclusion and opportunity: traditions of work amongst British Jews." In *Ethnicity at Work,* S. Wallman, ed. London: Macmillan.

Lever, A. 1988. "Capital, gender and skill: women homeworkers in rural Spain." *Feminist Review* 30: 3–24.

McClelland, K. 1989. "Some thoughts on masculinity and the 'representative artisan' in Britain, 1858–1880." *Gender and History* 1. 2: 164–77.

Marx, K. 1976. *Capital.* Volume 1. Harmondsworth: Penguin. Orig. pub., 1887.

Mayhew, H. 1985. *London Labour and the London Poor.* Harmondsworth: Penguin. Orig. ed., 1861.

Mies, Maria. 1982. *The Lace Makers of Narsapur: Indian Housewives Produce for the World Market.* London: Zed Press.

Mitter, S. 1986. "Industrial restructuring and manufacturing homework: immigrant women in the U.K. clothing industry." *Capital and Class* 27: 37–80.

Morokvasic, M. 1987. "Immigrants in the Parisian garment industry." *Work, Employment and Society* 1. 4: 441–62.

Morris, J. 1986. *Women Workers and the Sweated Trades.* London: Gower.

Phillips, A. and B. Taylor. 1980. "Sex and skill: notes towards a feminist economics." *Feminist Review* 6: 79–88.

Phizacklea, A. 1982. "Migrant women and wage labour: the case of West Indian women in Britain." In *Women, Work and the Labour Market,* J. West, ed, 309–25. London: Routledge.

Phizacklea, A. 1983. *One Way Ticket*. London: Routledge.

– 1987. "Minority women and economic restructuring: the case of Britain and the Federal Republic of Germany." *Work, Employment and Society* 1. 3: 99–116.

– 1990. *Unpacking the Fashion Industry*. London: Routledge.

Rainie, A.F. 1984. "Combined and uneven development in the clothing industry: the effects of competition on accumulation." *Capital and Class* 1: 141–56.

Rose, S.O. 1986. "'Gender at work': sex, class and industrial capitalism." *History Workshop* 21: 114–31.

Rowbotham, S. 1973. *Hidden from History: 300 Years of Women's Oppression and the Fight Against It*. London: Pluto Press.

Schmiechen, J.A. 1984. *Sweated Industries and Sweated Labour: The London Clothing Trades, 1860–1914*. London: Croom Helm.

Spelman, E.V. 1990. *Inessential Woman: Problems of Exclusion in Feminist Thought*. London: The Women's Press.

Stedman Jones, G. 1984. *Outcast London: A Study in the Relationship Between Classes in Victorian Society*. London: Penguin.

Strachey, R. 1978. *The Cause: A Short History of the Women's Movement in Great Britain*. London: Virago.

Stockley, Beth. 1988. Preface to *Woven Air: The Muslin and Kantha Tradition of Bangladesh*, Paul Bonaventura and Beth Stockley, eds. London: Trustees of the Whitechapel Art Gallery.

Taylor, B. 1983. *Eve and the New Jerusalem: Socialism and Feminism in the Nineteenth Century*. London: Virago.

Tilly, L.A. and J.W. Scott. 1987. Rev. ed. *Women, Work and Family*. London: Methuen. Orig. ed., 1978. New York: Holt, Rinehart and Winston.

Werbner, P. 1987. "Barefoot in Britain: anthropological research on Asian immigrants." *New Community* 14: 112.

Westwood, S. 1984. *All Day, Every Day: Factory and Family in the Making of Women's Lives*. London: Pluto Press.

7 Gender and Occupational Restructuring in Montreal in the 1970s

DAMARIS ROSE AND
PAUL VILLENEUVE

The massive growth of participation rates for women in the paid labour force and the accompanying feminization of the work force raise a series of questions with regard to the geography of work and employment. Indeed over the past decade an increasing number of authors have suggested that the spatial differentiation of labour-force characteristics, including gender, should be given renewed consideration as a key industrial location factor in many cases (Aydalot 1983; Clark 1981; Lipietz 1980; Massey 1984a; Raffestin and Bresso 1979; Scott 1982; Storper 1981). In this paper, which draws on our own research on the differentiation of labour-force characteristics and their relationship to gender and household characteristics within contemporary metropolitan regions, specifically those of Montreal and Quebec City,[1] we will explore some aspects of the relationship between the *feminization* of the paid work force and the changing *occupational* division of employment within different branches or sectors of the formal economy in the Montreal metropolitan area. Here we will deal only with paid work, although it is clear that, in order to understand the causal processes involved and the social impacts of recent changes in labour-force composition, the links between these developments and those in the domestic and community spheres must be examined (Armstrong and Armstrong 1985; Le Bourdais, Hamel, and Bernard 1987).

The feminization of the work force has been occurring while advanced capitalist economies are engaged in fundamental processes of capital restructuring and technological change, entailing, variously and unevenly, trends toward deskilling, reskilling, and an increased level of polarization or bifurcation in the occupational division of labour. Our focus here is on the link between feminization and these tendencies. We first discuss the relation between current forms of economic restructuring and changes in the gender division of employment; we then empirically explore the issues of deskilling, reskilling, and, in particular, polarization in the occupational structures of the female and male labour forces in the Montreal metropolitan area, using special tabulations of 1971 and 1981 census data. Finally we take a brief look at the occupational division of the male and female labour forces in places of work at different distances from downtown Montreal in order to see how deskilling, reskilling, and polarization tendencies vary within metropolitan space. We conclude with some observations about whether these three variables are representative of what is happening to the structure of women's employment, with some hypotheses as to the future direction of occupational restructuring of women's paid work and, of course, with a call for further research.

THE CHANGING GENDER DIVISION OF EMPLOYMENT BY ECONOMIC SECTOR AND OCCUPATION: SOME CONCEPTUALIZATIONS

We may conveniently begin with a cursory look at the sectoral structure by sex of the Montreal metropolitan area labour force in 1971 and 1981. Our sectoral categorization (Table 1) is derived, with certain modifications, from Simmie's work (1983), which is based on a critique of simplistic notions of postindustrialism, and in which he makes use of the categories of production, circulation, distribution, exchange, and consumption. Although doubts about data reliability forced us to use heavily aggregrated categories for our special tabulations of census data, our classification is nonetheless able to isolate key sectors of recent metropolitan development such as finance and services to business management.[2] Table 1 reveals trends broadly similar to those observed in Canada as a whole (Armstrong 1984) and in countries such as the United Kingdom (Lewis 1984), with both a marked feminization and a persistent segmentation by sex. The total female labour force has increased by more than 60 per cent during the decade, while the male labour force has grown by less than 25 per cent. In certain sectors, such as transportation, utilities, and trade, services to business, and public administration, the female labour force has almost doubled. What has happened to segmentation by sex during this period? A simple segregation index shows that the unequal distribution of men

Table I
Sectoral Structure by Gender, Montreal Metropolitan Area

Sector	Men 1971 No.	%	Men 1981 No.	%	Women 1971 No.	%	Women 1981 No.	%
CIP: Primary, construction, capital-intensive manufacturing	133,625	20.9	160,220	20.1	16,570	4.9	27,855	5.0
LIM: Labour-intensive manufacturing	110,535	17.3	114,620	14.4	70,375	21.0	90,580	16.2
DIS: Transportation, utilities, and trade	177,010	27.7	221,190	27.8	60,985	18.2	113,740	20.4
FIN: Finance, insurance, and real estate	32,210	5.0	35,445	4.5	29,135	8.7	52,165	9.3
EHW: Education, health, and welfare services	49,840	7.8	69,655	8.7	82,295	24.6	130,480	23.3
LCP: Communications, recreation, and personal services	68,070	10.6	100,850	12.7	51,310	15.3	89,830	16.1
BUS: Service to business	24,885	3.9	42,045	5.3	12,170	3.7	28,265	5.1
ADM: Public administration and defence	43,810	6.8	51,790	6.5	12,060	3.6	25,640	4.6
Total	639,985	100.0	795,815	100.0	334,900	100.0	558,555	100.0

Source: All tables in this paper are derived from special tabulations produced for the authors by Statistics Canada

and women across the sectors has slightly diminished, dropping from 29 per cent in 1971 to 25 per cent in 1981. This is the percentage of men or women who would have to be redistributed across sectors to arrive at no sectoral segregation; and while it may appear low, this is a significant figure given the high degree of sectoral aggregation used.[3] More detailed sectoral breakdowns would show more striking patterns of segregation, as research by Hanson and Pratt (1988) indicates.

Lewis (1984, 47) has argued that the feminization of the waged labour force in advanced capitalist economies has made the relationship between the sexual division of labour and processes of industrial reorganization more evident. Consequently it has become essential to analyze the sexual division of employment in order to gain an adequate understanding of the geography of contemporary economic restructuring (Massey 1984b, 11–12). In the relevant literature there is a strong emphasis on technological change as a primary mediating factor in recent rounds of employment restructuring (e.g. Bradbury 1985; Massey and Meegan 1982) and on the profound transformations this change could wreak in the skill structure of the work force (Kaplinsky 1984; Newton 1984), particularly with regards to women (e.g., Baran and Teegarden 1987; Menzies 1985). Yet there is no consensus as to the overall direction of technologically induced occupational change; rather, it would appear that while women in some sectors of the labour force do and will face the prospect of deskilling or the spectre of technological unemployment, the introduction of new technologies may open up opportunities of advancement for those more highly skilled or supervisory positions for other groups of women.

If we take the view that technology is not simply new hardware or machinery (cf. Freeman 1984, 23) but that which includes the "knowledge about the individual elements combined into a coherent labour process" (Toft-Jensen et al. 1983, 100), then technological change becomes a mediating factor instead of an independent variable in the analysis of changing occupational structure. This view of technology has led a number of authors to examine the relation between technological change and the skill structure of the work force. Here we encounter at least three apparently contradictory hypotheses that we will label, for the sake of brevity, as those concerning 1) reskilling; 2) deskilling; and 3) concomitant reskilling and deskilling, or what is often called polarization (for an overview of the literature, see Bernier 1984). We view technological change as a major contributing factor to all of these processes.

For some authors skill levels are almost necessarily raised by technological change (Aydalot 1983; Molle 1983); this is the *reskilling* hypothesis. For other authors the opposite perspective prevails: new technologies are seen to go hand in hand with a *deskilling* of the labour force (Massey and Meegan 1982; Hyman and Price 1983). And for still others it is a *polarization* in the occupational structure that is the most dominant characteristic of the intense capital restructuring and technological change experienced since the crisis of the early 1970s (Lipietz 1980; Noyelle and Stanback 1984). These points of view in reality only compete for the researcher who is concerned with change at a very aggregate scale. Most authors would probably now agree that different industries and sectors have followed different paths of restructuring, creating important variations in how the labour force is being reshaped (Bradbury 1989; Kuhn and Bluestone 1987; Massey 1984b; Rubery 1988). Aggregate patterns in a particular place will thus largely depend on the mix of economic sectors in that place, as we will discover for the case of Montreal. With this comment in mind we now will briefly review the three perspectives on the relation between technological change and skill structure.

The *reskilling* hypothesis is often based on the truism that overall levels of schooling have been increasing steadily, among the population as a whole but even more notably among women, in advanced economies. To this general point are added more specific arguments, according to which automation, along with new possibilities for control over information and over the production and distribution of goods, has decreased the proportion of production workers and increased the proportion of managers, professionals, and technicians in the work force. The same tendency towards a relative decrease of low-skilled jobs and an increase of skilled jobs has been noted in certain information processing industries, such as insurance (Baran and Teegarden 1987; Menzies 1982). For some authors this tendency is directly traceable to the introduction of microelectronics in both production and office work (Molle 1983).

The well-known *deskilling* thesis argues that it is erroneous to limit oneself to the study of changes in official schooling levels or occupational levels of the population. Rather it focuses on job content, whatever the name-tag of the job (Bernier 1984, 143). For instance even professionals and technicians could experience a simplification and a routinization of their work. The data available to us for present purposes do not enable us to examine such a hypothesis, but we will bear in mind throughout that an increase, or no change, in the number of jobs at a given occupational level, as indicated by census categories, can conceal deskilling in terms of job content.

The evidence for the *polarization* hypothesis is slightly less fragmentary and the process appears better theorized. A number of authors consider the increasing functional and spatial division of labour between control functions and routine tasks to be at the heart of polarization tendencies in the occupational structure of certain sectors of the economy (Lewis 1984; cf. Lipietz 1980, 7). With the spread of computerization in labour processes, the simultaneous concentration of control and fragmentation of manual operations can be furthered, as, for example, in the case of telephone operators (Menzies 1982, 1985); but microcomputers may also allow a new reintegration of previously fragmented tasks in certain types of professional, technical, or high-level white-collar work.

Polarization processes seem to be developing, then, in the occupational structures of certain service sectors as well as in manufacturing (for discussions of this literature, see Bednarzik 1988; Kuhn and Bluestone 1987; Myles et al. 1988). At one end of the occupational hierarchy the number of high-level managers, professionals, and technicians is increasing, especially in producer services, such as finance, and services to business management, although apparent professionalization in the latter sector should not be exaggerated, since many of these upper-level workers, especially women, are "independent contractors" who work on an on-call basis and often under insecure conditions (Christopherson 1988, 12). At the same time, these sectors require a large and low-waged support staff, ranging from data-entry clerks to office cleaners and security guards.

At the same time we are witnessing the secular trend of expanding new consumer service activities, which are replacing tasks formerly accomplished through unpaid labour in the domestic sphere, such as eating services (e.g., Castells 1976, 604). This trend is related to the feminization of the work force and the spectacular growth in numbers of two-earner households (Smith 1984; Villeneuve and Viaud 1987). Such trends may be generating another form of polarization not linked to technological change: the proliferation of new, low-wage service jobs primarily taken up by women (Armstrong 1984; Smith 1984).

In major urban centres of advanced tertiary employment, we see these two trends coming together in the same geographical space, because the expansion of low-wage jobs in the realm of personal services is also in part a consequence of the major (growing) concentrations of highly paid managers and professionals, the "new urban elite" (cf. Hutton and Ley 1987; Smith 1986) whose "conspicuous consumption" habits require servicing (Sassen-Koob 1984). Some authors have thus argued that polarization tendencies in the tertiary occupational structure may thus be greatest in those "global" or "national" cities where major corporate headquarters are located (Smith and Feagin 1987); moreover, in such cities overall polarization may be reinforced if a large sector of immigrant labour working in low-waged, labour-intensive manufacturing is also present (Sassen 1988). As we will see, although Montreal is not really in the league of international cities, the recent evolution of its economic sectoral structure as well as its historical legacy of low-waged manufacturing have been quite conducive to trends in certain forms of polarization; and feminization processes have made important contributions to these trends.

WOMEN AND OCCUPATIONAL RESTRUCTURING: THE CASE OF MONTREAL IN THE 1970S

Our empirical study of the transformations in the industrial and occupational structures of the Montreal labour force is only indirectly a study of the restructuring of skills, since we are restricted to information based on job titles as classified by the census, regardless of hidden changes in job content.

Our seven occupational categories (Table 2) are based on the complementary notions of control and skill, as defined in the detailed descriptions of job titles given in the *Canadian Dictionary of Occupations*. As far as is possible, given the limitations of even these detailed (four-digit) census classifications, we distinguish control over others (managers and supervisors or forepersons) from control built directly on skills, whether the latter are grounded in knowledge (professionals) or in know-how (upper-level white-collar workers, including technicians, and skilled blue-collar workers). Our categories are thus related to each other through asymmetrical relations of power based on the capacity to exercise control and skills; they are similar to, although not as detailed as, those used in a more recent study by Drouilly and Brunelle (1988), but they critically analyze more conventional ways of grouping and labelling occupational categories that rely on the notion of status (Drouilly and Brunelle 1988). Thus ours seem to be more appropriate than the latter for exploring questions of change in skill and in degrees of polarization. (It should also be mentioned that the ranking of the seven categories is based upon the mean employment incomes [of both sexes combined] for each category).

We now turn to an analysis of the evolution of the occupational structure for the Montreal Census Metropolitan Area (CMA) between 1971 and 1981. First, we consider the aggregate occupational structure (Table 2). While we see a sometimes impressive increase in *absolute* numbers across all occupations, small changes in the column-wise

Table 2
Occupational Structure, Montreal Metropolitan Area

	1971		1981	
Category	No.	%	No.	%
MAN: Managers	32,815	3.5	80,110	5.9
PRO: Professionals	121,405	12.7	170,170	12.6
SUP: Supervisors	94,705	9.9	118,145	8.8
SKW: Upper-level white-collar	180,570	18.9	279,270	20.7
SKB: Skilled blue-collar	117,665	12.4	146,240	10.9
UKW: Lower-level white-collar, sales and services	278,650	29.2	384,670	28.5
UKB: Semi- and unskilled blue-collar in production	127,570	13.4	169,455	12.6
Total	953,375	100.0	1,348,055	100.0

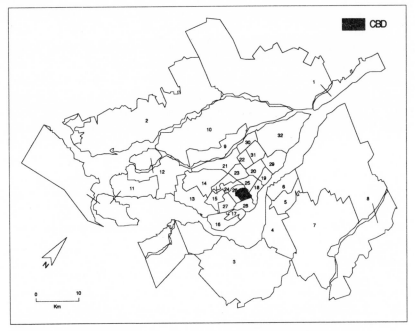

Figure 1
The Montreal Metropolitan Area Divided into 32 Zones

percentages are also significant insofar as they are structural and based on large numbers. As will be seen, some of our data lend partial support to the polarization hypothesis and illustrate the salience of gender in polarization processes. Our results are in some ways consistent with the reskilling hypothesis, too, although our inability to consider changes in job content means that great caution should be used in interpreting results that suggest an upward movement in the occupational structure, especially for women.

For the overall occupational structure (all sectors and both sexes), there is some reskilling but no clear evidence of increasing polarization during the 1970s, if the latter is defined simply as a larger relative gain in the percentages of workers in upper and lower categories than that occurring in the middle of the occupational spectrum (Table 2). At this highly aggregated level, very general tendencies can be noted: we see slight percentage decreases in lower-level occupations; more pronounced decrease at the level of supervisors and forepersons and also at the skilled blue-collar level; practically no change at the professional level; and marked increases at the managerial and upper white-collar and technical levels. Indeed the data mostly reflect the evolution of Montreal as a centre of administration and control for the Quebec economy (Cossette 1982; Polèse 1988) and the massive growth of the state apparatus and parapublic sector in Quebec in the 1970s, which generated large numbers of upper-level and intermediate white-collar jobs. Yet the city's economy also relies increasingly on tourism and other sectors that generate low-level service work (Lamonde and Polèse 1984) while, in spite of deindustrialization, both skilled and unskilled manufacturing jobs still employ a higher proportion of the work force than found in many major cities that function as centres of advanced services. Although the two lowest-ranked occupational groups decreased slightly in relative terms over the decade, and thus could not be said to have contributed to increasing the polarization in the occupational structure, the large absolute increase in lower-level, white-collar sales and service employment combined with the relative stability of low-skilled manufacturing do indicate that Montreal's occupational structure in 1981 was a fairly bifurcated one.

Disaggregating the overall structure by gender (Table 3), we first notice the strong occupational segregation between the sexes. Women are significantly overrepresented in the two white-collar categories and they are strongly underrepresented in the skilled blue-collar category. They are, however, almost at par with men professionals, although their employment income in this category was still only 63 per cent of that of men in 1980 (Rose 1987, 215). We also observe that percentages of women in higher-level categories tend generally to be lower than those of male workers, although the gap closed somewhat over the 1970s due to an increase in the number and proportion of female managers. The latter is a real and significant trend in Quebec, but

Table 3
Occupational Structure by Gender, Montreal Metropolitan Area

Category	Men				Women			
	1971		1981		1971		1981	
	No.	%	No.	%	No.	%	No.	%
Managers	29,185	4.7	62,875	8.0	3,630	1.1	17,235	3.1
Professionals	83,565	13.4	102,070	12.9	37,845	11.5	68,100	12.2
Supervisors	78,675	12.6	89,340	11.3	16,025	4.8	28,805	5.2
Upper-level white-collar	80,035	12.9	106,080	13.4	100,535	30.4	173,195	31.0
Skilled blue-collar	106,510	17.1	133,050	16.8	11,155	3.4	13,185	2.4
Lower-level white-collar	161,465	25.9	191,640	24.2	117,175	35.5	193,025	34.6
Semi- and unskilled blue-collar	83,625	13.4	105,450	13.4	43,940	13.3	64,000	11.5
Total[1]	623,060	100.0	790,505	100.0	330,305	100.0	557,545	100.0

[1] The slightly different totals between occupations and sectors (Table 1) are due to identification difficulties in census taking.

numerically the representation of women in management is still very limited (see Gold in this volume); in 1981, they were still outnumbered by men in a ratio of 3.6 to one. Due to the migration of head offices of manufacturing and finance companies to Toronto in the 1970s, the proportions of male managers and professionals in Montreal remained virtually stagnant, making the increase for women seem particularly impressive. As Montreal has become less of a high level corporate control centre for the whole of Canada, women have come to form a higher proportion of its managers and professionals, because women are concentrated in less senior positions (Drouilly and Brunelle 1988) as well as in sectors in which Montreal maintains a specialized function for the province of Quebec – notably education, health, and welfare (Rose 1987).

More broadly, we may imagine that what appears as reskilling in the female work force in major centres of office employment like Montreal is in fact a function of new spatial divisions of labour in white-collar work, divisions that operate at various geographical scales and effectively create a "metropolis-hinterland" phenomenon of spatial polarization of office work (cf. Mitter 1988) not just *between* men and women but, and perhaps increasingly so, *within* the female labour force. With the transfer of some routine office work to suburban "back offices" (Huang 1989; Nelson 1986) or to smaller towns in the regions (as was the case for federal income-tax processing, for instance), and with the use (by insurance companies and magazine subscriptions, for example) of satellite technology that allows data entry to be done more cheaply in the southern U.S. or offshore (Cohen 1987, 28–32), it is not surprising that a higher proportion of the white-collar work that remains located in a major city is likely to consist of higher-level or relatively specialized jobs.

The data in Table 3 suggest that the female occupational structure was, overall, less polarized than that of men in both census years; this is because women are overwhelmingly concentrated in middle-level, white-collar jobs, and also because reskilling did seem to be the dominant overall trend of the 1970s in the female employment struc-ture – which indicates that the gains made as a result of the women's movement were not illusory ones. The expansion of the public and parapublic sectors of education, health, and welfare in Montreal was a major factor that enabled increasing numbers of women to gain access to middle- and upper-level positions in the 1970s, although they also made gains in the professional and managerial job categories in the burgeoning producer-services sector. Yet at the same time, women's employment in low-skilled manufacturing jobs (most notably in the clothing industry) has been remarkably tenacious (Lamonde et al.

1988), and even more so if we include informal homeworking in this sector (estimated at 22,000 jobs in 1983, according to a study cited in Lamonde and Polèse 1984). Indeed, not unlike the situation in more truly global cities like London or Los Angeles (see Mitter 1988; Sassen 1988), recourse to a continuing supply of immigrant women in precarious circumstances and thus with little or no bargaining power has helped some segments of Montreal's clothing industry to refit and even prosper in the face of increased foreign competition (Mummé 1983; cf. Morokvasic et al. 1986). Montreal's sectoral structure may well have been conducive to an increasing polarization of the female work force in the 1980s and beyond, although in the 1970s overall polarization tendencies within Montreal's female work force were muted by an expansion of skilled white-collar and technical jobs about equal to that of less skilled, white-collar sales and service jobs. Within certain sectors, however, there *is* evidence of increasing occupational polarization of the female work force during the 1970s. As we will see, the trend is most clear in services to business, in finance, and in labour-intensive manufacturing (cf. Sassen-Koob, 1984; 1986).

We should point out that while it is important to examine separately the polarization levels of the male and female work forces, these should not blind us to the fact that, overall, the persistence of labour market segmentation by gender – combined with the somewhat dualistic sectoral structure of major centres of advanced services – means that even those advanced sectors that are increasingly open to employing women as managers and professionals (finance and business services) retain occupational hierarchies strongly polarized, along gender lines, between male professionals and managers and female low-level office and service workers.

We now turn to further disaggregation of occupations by economic sector.[4] Figure 2 presents the detailed occupational distributions in each sector.

Production. Our classification divides the production sector into two: production sector I (CIP) includes *capital-intensive manufacturing, primary activities* (such as head offices of mining firms), and *construction*; and production sector II (LIM) comprises *labour-intensive manufacturing,* in which female employment is typically more important.

Production sector I is the only area wherein female workers can be said to have a higher occupational level than that of men. This is because the few women in this sector are strongly concentrated in upper-level white-collar jobs, while men are almost as strongly concentrated in skilled blue-collar jobs. Here the polarization is clear for both female and male workers: the proportions in extreme categories

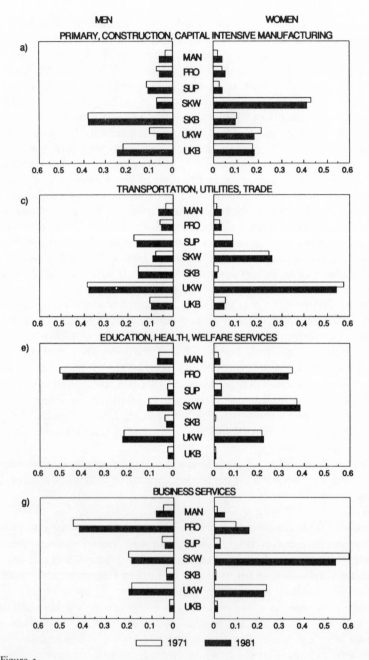

Figure 2
The Occupational Structure of the Metropolitan Labour Force, 1971 and 1981[1]
[1] Occupational categories are defined in Table 2.

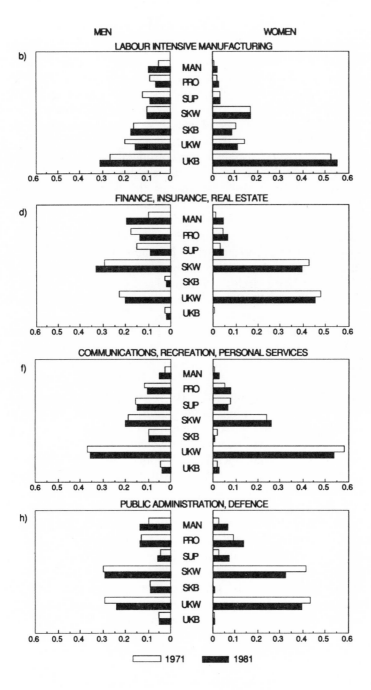

MEN   WOMEN

b) LABOUR INTENSIVE MANUFACTURING

d) FINANCE, INSURANCE, REAL ESTATE

f) COMMUNICATIONS, RECREATION, PERSONAL SERVICES

h) PUBLIC ADMINISTRATION, DEFENCE

1971   1981

increased over the decade, notably for men in low-skilled blue-collar work and for women in high-skilled work. The manufacturing industries included in the sector have undergone quite pronounced capital restructuring and technological change during this period. Take for instance the transportation equipment industry, well represented in the Montreal CMA. It expanded considerably during the seventies and more than doubled its number of workers, with almost all of this growth taking place in suburban industrial areas.

Production sector II also exhibits considerable, and increasing, polarization. In fact this sector is the most polarized of all, both for women and for men, mainly because of the high and increasing proportions in semi- and unskilled blue-collar jobs. Women form almost 45 per cent of the employees in this sector, and they are heavily concentrated in low-skilled blue-collar work. The gender effect on polarization could not be clearer than it is here: male managers still outnumbered their female counterparts by 6 to one in 1981; total office jobs showed a relative decline; skilled blue-collar jobs increasingly went to men; and low-skilled production jobs were occupied by women in a proportion of 1.5 to 1. Production sector II industries are more centrally located than those in sector I, in part because of lower space requirements and in part because of their continued links to local pools of immigrant female labour, as in the case of the clothing industry.

Distribution. The distribution sector (DIS) comprises *wholesale and retail trade, transportation,* and *public utilities.* Probably because of the relatively small impact of technological change in this sector in the period under study, we find relatively little change in the occupational structure, except for slight increases in upper categories and slight decreases in lower categories (both of which are a little more pronounced for women than for men).

Exchange. Exchange activities (FIN) essentially take place in the sectors of *finance, insurance,* and *real estate* and facilitate the exchange of land, labour, capital, raw materials, goods, and services through the medium of money (Simmie 1983, 60). Because this is so heavily a white-collar sector, upper-level categories, especially within the male work force, are relatively important. This is one sector that has undergone marked computerization, and while we cannot establish a direct relation between this fact and the occupational shifts observed here, the very high proportions of male managers, on the one hand, and female lower-level white-collar workers, on the other hand, are quite striking. In 1971 men outnumbered women in this sector, but ten years later it contained substantially more women than men (Table 1). So despite apparent occupational upgrading for women in this sector (with the three top categories all showing increases), in absolute terms it employed many more lower-level white-collar women in 1981 than in 1971.

Services. If we consider first service sector 1, *education, health* and *welfare* services (EHW), the great majority of which are public or parapublic services in Canada, we see that this component of the sector has the most upwardly skewed occupational distribution for both men and women, but most especially for the latter. This is not surprising in view of the strong professionalization of these fields. Polarization did not appear to have increased, and it is interesting to note that the proportion of female lower-level white-collar workers has not decreased as it has in distribution and exchange.

Service sector II (LCP) comprises a whole range of *consumer* and *personal* services, delivered privately in the main, from communications to hotels and restaurants. This is not a sector that has witnessed heavy overall technological change, except within communications. The restructuring in this sector is slightly more pronounced for women than it is for men, partly because of positive changes both at the managerial and professional levels and at the semi- and unskilled blue-collar level. Indeed among the service sectors, this division has the highest share of semi- and unskilled workers, especially female.

Service sector III (BUS) includes the whole range of services offered to *business management*. It is a small but fast growing sector in Montreal and is concentrated in the downtown area, where employment for this sector increased by over 55 per cent between 1971 and 1981 (Polèse and Lamonde 1984, 34). This sector is second only to education, health, and welfare in terms of its high proportions of workers, both male and female, in the upper categories. It also shows a quite clear polarization in the female occupational structure, which increased during the 1970s with a considerable relative drop in upper-level white-collar staff; reskilling is also in evidence, and relative increases at the top levels are more pronounced than the decrease at the semi- and low-skilled white-collar level. Among men the two lower-level categories (these comprise mostly janitorial and cleaning services and security guard positions) increased markedly over the period.

Public Administration. The public administration sector (ADM) consists of direct *federal, provincial,* and *local government* employment. Blue-collar jobs excepted, it is here that we find the most regularly shaped occupational hierarchy. Here, in contrast to most other sectors, supervisory jobs were on the increase in the 1970s, and there was a relative decrease in upper-level white-collar work. It is in this sector that affirmative action programs have had the most pronounced effects on job opportunities for women in the province of Quebec (Maroney 1983); this is reflected in the fact that the proportion of women professionals working in this sector is the same as of men, while the proportion of supervisors is larger for women than for men. The gap at the managerial level is still, however, extremely wide.

At least three general observations can be drawn from the data presented so far. First, each sector of the Montreal economy has evolved in a specific way with respect to its occupational structure during the decade under study, which underlines the point made earlier about the danger of excessively glib one-word characterizations of dominant tendencies in restructuring of occupational hierarchies. Although, in our analysis, the economy was partitioned into only eight sectors, the categorization scheme adopted has been able to indicate, in a simple, descriptive way, the particular evolution of each sector as well as the relative position of women and men within each sector.

Secondly, our data suggest that most sectors of the Montreal economy underwent an apparent occupational reskilling, modulated differently for women and men, during the 1970s. Only qualitative, detailed, sector-by-sector studies, however, will be able to discern to what extent this tendency reflects an upgrading in tasks performed and to what extent an upward reclassification on paper, the latter of which goes hand-in-hand with the tendency of employers to demand higher qualifications and more previous training and experience for most non-manual employment.

Thirdly, in certain economic sectors we found that there may have been an increase in occupational polarization over the decade both between and within each of the male and female labour forces, but slightly more pronounced within the latter. It could be argued that this larger increase in the polarization of the female occupational structure is due to the more recent entry of women into the labour force, combined with the persistence of labour-market discrimination, especially against part-time workers and recent immigrants from minority cultures (see Lamotte 1985). Yet the fact that certain sectors also show increases in the level of polarization in the male occupational structure might indicate that capital restructuring has also played an important role in these processes.

SPATIAL ASPECTS OF OCCUPATIONAL RESKILLING AND POLARIZATION

Finally, to give some indication of the spatial patterning of apparent reskilling and polarization tendencies in different parts of the Montreal CMA, we analyzed the occupational structure of the work force by sex, for all economic sectors combined, and at place of employment in each of the zones shown in Figure 1.[5] We examined the values of median occupational ranks, and quartile deviations in occupational ranks, at different distances from the centre in 1981. We also graphed the amount of change in these statistics against distance from Montreal's Central Business District or CBD (Figures 3 to 6).

For women median levels were highest in the CBD, where there was a rapid increase from 1971 to 1981 (which indicates reskilling), and in certain middle-class suburbs (where the main sources of employment are in education, health, and welfare institutions and in services to the resident population). Indeed our place-of-employment data indicate that professional jobs for women are quite dispersed in metropolitan space, although there is a major concentration in the central part of the Island of Montreal. At the same time, very low median levels were still found for 1981 in centrally located districts where clothing and other low-wage manufacturing employment predominate. For men the median levels are highest by far in the CBD and decrease from the centre in a fairly regular pattern, except for a few suburban places of employment, such as zones 7 (St-Hubert, St-Bruno) and 11 (Kirkland, Ste-Anne-de-Bellevue, etc.) where the headquarters of some major high-technology corporations are located.

Figure 3 c) shows a clear relationship between distance and change in women's median occupational level, and Figure 4 c) a much weaker relationship for the case of men. The median occupational level of women actually *decreased* in nearly half of our thirty-two areas, with the biggest decrease in zone 10 (the newer part of the municipality of Laval), to which there has been some movement of back offices partly on the basis of a large female labour pool (*Les Affaires* 1982), and most of these decreases are in locations that are more than ten kilometres away from the city centre. In the case of men median occupational levels decreased in only seven zones, while instances of increase are generally of a lower magnitude than for women. These tendencies can be fully understood only with detailed reference to the sectoral mix of each zone, but in general the figures confirm that overall upward mobility has been more pronounced in centrally located economic sectors, such as public administration and financial and business services.

We turn now to an analysis of the degree of *dispersion* in the occupational structure of each zone (Figures 5 and 6). As one might expect, the female employment structure in 1981 shows the largest quartile deviations to occur in those central districts that have both traditional industrial employment and large educational, social service, and medical institutions. In the CBD itself (zone 26), dispersion levels

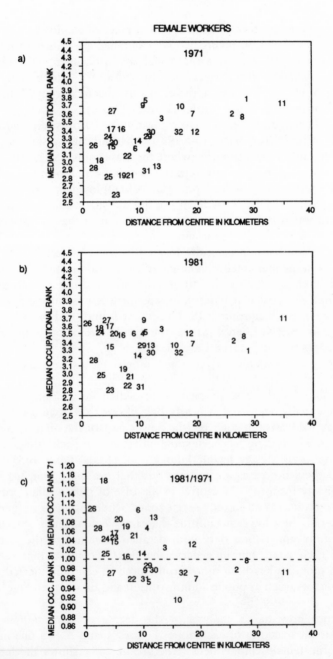

Figure 3
Median Occupational Ranks of Female Work Force in Metropolitan Montreal,[1] 1971,
1981, and ratio of 1981/1971
[1] Thirty-two zones, numbered as in Figure 1.

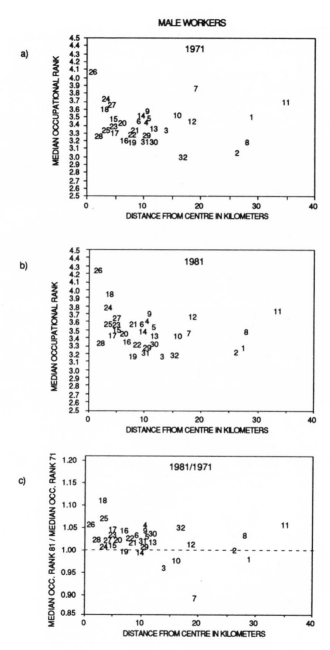

Figure 4
Median Occupational Ranks of Male Work Force in Metropolitan Montreal,[1] 1971,
1981, and ratio of 1981/1971
[1] Thirty-two zones, numbered as in Figure 1.

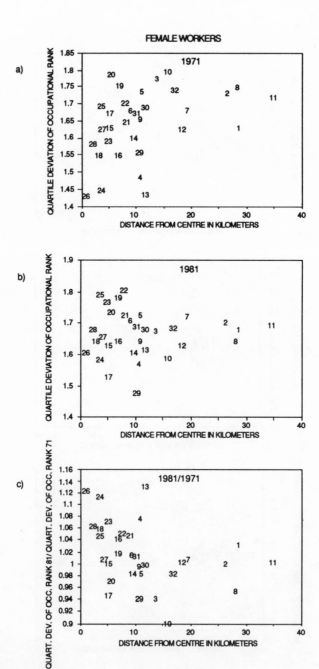

Figure 5
Quartile Deviations in Occupational Ranks of Figure 3[1]
[1] Thirty-two zones, numbered as in Figure 1.

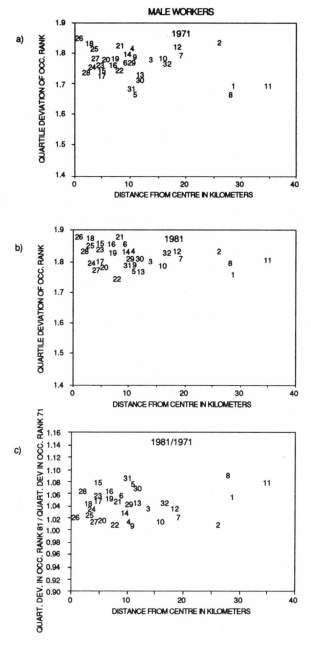

Figure 6
Quartile Deviations in Occupational Ranks of Figure 4[1]

[1] Thirty-two zones, numbered as in Figure 1.

were not very high but increased rapidly over the decade. While these are place-of-employment data, they have clear parallels in residential locations: in some of Montreal's inner-city neighbourhoods women professionals are an important fraction of the new middle-class, while working-class women, especially immigrants, maintain a strong presence (Rose 1987; Rose and Chicoine 1991). With respect to women's employment, quartile deviations show a slight tendency to decrease from the centre. For men the patterns are far less clear, probably because the Montreal region in the 1970s saw far more decentralization of middle-level jobs (notably in the manufacturing sector) for men than for women.

The graphs of changes in the quartile deviations of occupational ranks from 1971 to 1981 show contrasting patterns for women and for men. In the case of women changes in quartile deviation vary greatly between districts located within a fifteen-kilometre radius from the city centre, although there is a general tendency for this statistic to increase less rapidly in the suburbs than in the centre. For men quartile deviations increased slightly everywhere, whereas for women they increased most notably in the CBD, in surrounding inner-city districts with a great deal of female employment, and in the burgeoning industrial-commercial suburb of Dorval (zone 13, home to Montreal's busiest airport). The deviations decreased in some older industrial areas, possibly due to the disappearance of production jobs in the wake of deindustrialization. No such differentiation pertains in the case of men, for whom increases in quartile deviations are small and vary little across metropolitan space.

CONCLUSION: FUTURE TRENDS AND RESEARCH DIRECTIONS

The foregoing analysis is but a very modest step toward deepening our understanding of changes in the gender composition of the occupational structure brought about in the recent period of capital restructuring; it is also a modest step toward documenting the spatial patterning of these changes within a metropolitan area. Yet in spite of the crude measurement techniques used and the limitations of our extensive research method (see Sayer 1984, 219–28), this analysis of transformations in the occupational structure of the Montreal CMA labour force in the 1970s lends some credence to the polarization hypothesis for certain economic sectors and in certain areas, although not for the occupational structure of the metropolitan region as a whole. It is quite probable that in the 1970s, due to the massive growth of public and parapublic sector employment (cf. Myles 1988), trends toward occupational polarization resulting from the expansion of both very high-level and very low-level jobs in some branches of the economy (notably in the private sector) were greatly muted in Canadian cities, especially in the province of Quebec. Our study also sheds light on the role played by gender in polarization processes. We have found that simple measures of occupational levels and occupational disparities, when gender and location are controlled, can tell us a great deal about shifting occupational hierarchies at the metropolitan level and perhaps in local labour markets – although in the present paper we have barely touched upon the latter scale of analysis.

With respect to the hypothesis of reskilling, although there are signs of this trend in some sectors and notably within the female work force, we are cautious about the type of inferences that can be drawn on the basis of Statistics Canada's occupational categories, which are inevitably too inflexible to provide an accurate indication of changes in job content during a time of rapid technological change in the waged workplace.

It must also be remembered that our analysis is limited to the 1970s; unfortunately we do not have access to comparable data from the 1986 census that could allow us to systematically extend our Montreal study into the mid-1980s. It is worth noting, however, that a recent Statistics Canada countrywide study covering the period from 1981 to 1986 – thus including the recession of the early 1980s, which seems to have provided the impetus for significant occupational and wage-rate restructuring within various economic sectors – shows that over this period there was clearly a more rapid growth both of professional and managerial jobs, on the one hand, and of sales and service jobs, on the other, compared to that of intermediate-level jobs. This trend has so far affected both the male and the female work forces (Myles et al. 1988). In our view, however, it could eventually affect the female occupational structure to a greater extent, especially if middle-level jobs in the public and parapublic sectors are further cut back by neoconservative policies – policies that also effectively encourage deskilling and the growth of low-waged service work in an unbridled free enterprise culture through the contracting-out of some public services, and through the commercialization or "informalization" of others (e.g., child day-care). As Myles (1988) points out, polarization tendencies (whether in occupational structures or in wage levels), far from being, as some authors claim, the inevitable result of sectoral shifts in a postindustrial society, are to a large extent the product of political decisions. This probably explains in part why the process of polarization in the female work force in the USA (Power 1988) seems to be more advanced than in Canada.

Should future growth in low-level white-collar sales and service occupations surpass that in intermediate white-collar positions, and should the proportions of women in managerial and professional occupations in the advanced tertiary sectors continue to increase, we can expect latent polarization trends in the female work force to become more evident in the near future in cities like Montreal. The limited published data for detailed occupations in 1981 and 1986 at the CMA level (Statistics Canada, Census of 1986, cat. 93-156, Table 16) provide hints that this polarization is happening. On the one hand we see women making major inroads, relatively and absolutely, in the previously male-dominated positions of sales and advertising directors and accounting and auditing, for instance; on the other hand women in Montreal also increased their share of routine sales positions. Moreover the impact of technological change on office work that is evident for these years is perhaps presaged by the massive absolute decline, from 1981 to 1986, in the female-dominated occupation of bookkeeping in Montreal (even in the then-booming city of Toronto, numbers in this occupation were stagnant) and a spectacular surge in the number of electronic data-processing operators.

In this paper we were not able to relate polarization in the occupational structure of different economic sectors to employment incomes, as has been attempted in some recent American work (Bednarzik 1988; Kuhn and Bluestone 1987; Sassen-Koob 1984; 1986) as well as in the Statistics Canada study (Myles et al. 1988). The latter analysis, notably, shows that the share of higher-paying jobs increased in public administration, transportation, communications, natural resources, and capital-intensive manufacturing, while that of low-paying jobs increased in the already low-wage sector of consumer services, as well as in construction. Given the importance of all of these sectors (except resources) in Montreal's economy, these findings point to a likely increase in wage polarization within both the male and the female labour forces in the region.

Moreover the Statistics Canada study points to an increased bifurcation, Canada-wide, of hourly wages between younger (i.e., under 35) and older workers and not only (as one might have expected) in the consumer services sector. Again this trend affects both men and women, but it must be remembered that women still earn much less than men, *ceteribus paribus*. Thus if we wish to consider the potential impacts of such processes on the living situations of the households in which individual wage earners live, we can easily imagine, for instance, the particularly acute situation faced by a household, the principal earner of which is a young woman working at the lower pole of the occupational distribution.

More broadly we believe that if researchers are ultimately concerned with the social implications of reskilling and polarization processes in major cities (with respect to housing and neighbourhood change, upward and downward social mobility), it will be necessary to consider the impacts of these processes that are reshaping the structure of the waged work force on the incomes, income prospects, and living standards of households – households composed of increasingly diverse permutations: single women, single men, mother-led families, two-earner couples, and unrelated people sharing accommodation for extended periods of time are some examples. Research on gender and employment will thus have to be linked not only to space but to *place* through studies of how people in different employment situations form themselves into households in different parts of the city (see Rose 1987; Séguin and Villeneuve 1987). We need in turn to explore how household and domestic situations, in conjunction with the facilities and services offered in the local area (Fincher 1988; Rose and Chicoine 1988), shape not only an individual's access to a given array of jobs but also the types of local jobs offered and determine to which segments of the population they are made available.

NOTES

Both authors are equally responsible for this paper. We are grateful to all those who made critical comments on earlier versions and to Marc Miller and René Morency for research assistance. This work was supported by the Social Sciences and Humanities Research Council of Canada (grant no. 410-84-0528).

1 This research has given rise to a number of publications, each of which illustrates the use that can be made of special tabulations of census data that link these characteristics (see Rose 1987; Rose and Villeneuve 1988; Séguin and Villeneuve 1987; Villeneuve 1989; Villeneuve and Rose 1986, 1988; Villeneuve and Viaud 1987).

2 Our sectoral classification is based on the premise that, while the now commonplace categorization of the economy into primary, secondary, and tertiary sectors (to which some would add quaternary) might have some use in general description, it lacks theoretical rigour. It appears to be more fruitful to focus on the five general functions of an advanced economy: production, exchange, distribution, consumption, and control (Simmie 1983, 60). Production thus includes mining, construction, and manufacturing, whether in the public or in the private sector. Exchange is essentially made up of finance, insurance, and real estate sectors. Distribution of commodities includes transportation, storage, utilities, and trade.

Consumption services comprise services to business; "human services," such as education, health, and welfare; and a whole array of personal services, such as communications services, culture and recreation, food, and lodging. The control function identifies the role of the state in regulating the economy as a whole. It includes public administration and defence, that is, the departments and ministries that compose the state proper, but it excludes publicly owned corporations operating in all of the above sectors. Inevitably certain misclassifications and ambiguities arise in the process of trying to operationalize this schema using the census, because even though we can ask the census for custom-built categories, these would be limited by the basic 350 or so sectors of economic activities that are first used to classify the individual census returns. For instance a bus driver would be included in the transportation category even though he or she provides a service to people rather than moves goods, while communications increasingly involves the transmission of information as a commodity as well as functioning as a personal service. We have expanded these five basic functions to eight categories (this limit was imposed in order to keep our cross-tabulations manageable and statistically viable). First, to isolate the more labour-intensive from the more capital-intensive manufacturing, we split production into two parts. Second, to isolate services in which women are overrepresented, and those that are rapidly growing and apparently concentrating downtown, we divided consumption into three parts: privately delivered services to business, publicly provided social services, and (largely) privately delivered personal services. Our sectoral classification thus has a firm theoretical grounding, and is at the same time adapted to our specific empirical concerns and to the practical limitations of the census (for more details on our methodology, see Villeneuve and Rose 1986).

3 This segregation index, also called the dissimilarity index, is widely used to measure residential segregation in urban social geography, but nothing precludes its use for measuring dissimilarity in distributions other than geographical ones. It is obtained simply by first finding the percentage of women in each sector or occupation throughout the total female labour force and by doing the same for men. Differences between the percentage of women and the percentage of men in each sector or occupation are then summed in absolute terms, and the sum is divided by two. The index ranges from zero (no segregation) to 100 (total segregation if all men were in one sector and no women were in the same sector).

4 For the eight previously mentioned sectors of economic activity we were able to obtain cross-tabulations of seven occupational groups by sex. We have this material for the metropolitan area as a whole and for the thirty-two zones of residence that we have defined within it (Figure 1). We also obtained data on each zone of employment by sex and occupation and

by sex and economic sector (Villeneuve and Rose 1988); we did not, unfortunately, obtain the cross-tabulation of the two, although that would have been particularly interesting in relation to recent work on women and the changing spatial division of labour within metropolitan areas (see, for e.g., Huang 1989; Nelson 1986).

5 To summarize the occupational structure in each zone, we have computed two simple statistics: the median occupational rank and the quartile deviation. In order to obtain these statistics, we considered our occupational categories (lowest of which were aggregated) to have defined an ordinal scale, with rank of 5 for managers, 4 for professionals, 3 for supervisors, 2 for white- and blue-collar skilled workers, and 1 for semi- and unskilled white- and blue-collar workers. The quartile deviation is simply the third quartile minus the first quartile, divided by 2. It can be taken as an indicator of the dispersion in our occupational structures; for example an increase of this index during the seventies would indicate increased polarization. Figures 3 and 4 plot median occupational ranks and changes therein during the seventies against the distance of each of the thirty-two localities considered as places of employment from Montreal's city centre. Figures 5 and 6 show quartile deviations.

REFERENCES

Les Affaires. 1982. "Dossier économique: Laval," mai (supplément), 1–26.

Armstrong, P. 1984. Labour Pains: Women's Work in Crisis. Toronto: Women's Educational Press.

Armstrong, P. and H. Armstrong. 1985. "Political economy and the household: rejecting separate spheres." Studies in Political Economy 17: 167–77.

Aydalot, D. 1983. "La division spatiale du travail." In Espace et localisation: la redécouverte de l'espace dans la pensée scientifique de langue française, J.H.P. Paëlinck and A. Sallez, eds, 175–200. Paris: Economica.

Baran, B. and S. Teegarden. 1987. "Women's labor in the office of the future: a case study of the insurance industry." In Women, Households, and the Economy, L. Beneria and C.R. Stimpson, eds, 201–24, New Brunswick, NJ: Rutgers State University Press.

Bednarzik, R.W. 1988. "The 'quality' of U.S. jobs." Service Industries Journal 8: 127–35.

Bernier, C. 1984. "Nouvelles technologies: requalification ou déqualification du travail?" Interventions économiques 12: 137–52.

Bradbury, J. 1985. "Regional and industrial restructuring processes in the new international division of labour." Progress in Human Geography 9: 38–63.

– 1989. "Strategies in local communities to cope with industrial restructuring." In Social and Economic Change in Industrial Societies in the 1980s, B. Van der Knaap and G.J.R. Linge, eds, 167–84, London: Routledge.

Castells, M. 1976. "The service economy and postindustrial society: a sociological critique." *International Journal of Health Services* 6: 595–607.

Christopherson, S. 1988. Labor flexibility: implications for women workers. Paper presented to the annual conference of the Institute of British Geographers, Loughborough, January, Department of City and Regional Planning, Cornell University, Ithaca, NY.

Clark, G.L. 1981. "The employment relation and spatial division of labor: a hypothesis." *Annals of the Association of American Geographers* 71: 412–24.

Cohen, M.J. 1987. *Free Trade in Services: An Issue of Concern for Women.* Ottawa: Canadian Advisory Council on the Status of Women.

Cossette, A. 1982. *La tertiarisation de l'économie québécoise.* Chicoutimi: Gaëtan Morin.

Drouilly, P. and D. Brunelle. 1988. "Une évaluation critique de la classification socio-économique des professions." *Interventions économiques* 19: 185–202.

Fincher, R. 1989. "Class and gender relations in the local labor market and the local state." In *The Power of Geography: How Territory Shapes Social Life*, J. Wolch and M. Dear, eds, 93–117.

Freeman, C. 1984. "L'économie de la recherche et du développement." In *Les enjeux du progrès*, A. Cambrosio and R. Duchesne, eds, 21–84. Montréal: Presses de l'Université du Québec.

Hanson, S. and G. Pratt. 1988. "Spatial dimensions of the gender division of labor in a local labor market." *Urban Geography* 9: 155–79.

Huang, S. 1989. Office suburbanisation in Toronto: fragmentation, workforce composition and labour sheds. PhD diss., University of Toronto.

Hutton, T. and D. Ley. 1987. "Location, linkages and labor: the downtown complex of corporate activities in a medium size city, Vancouver, British Columbia." *Economic Geography* 63: 126–41.

Hyman, R. and R. Price. 1983. *The New Working Class? White-Collar Workers and their Organizations.* London: MacMillan.

Kaplinsky, R. 1984. *Automation: The Technology and Society.* Harlow, UK: Longmans.

Kuhn, S. and B. Bluestone. 1987. "Economic restructuring and the female labor market: the impact of industrial change on women." In *Women, Households and the Economy*, L. Beneria and C.R. Stimpson, eds, 3–32. New Brunswick, NJ: Rutgers State University Press.

Lamonde, P. and M. Polèse. 1984. "L'évolution de la structure économique de Montréal, 1971–81: désindustrialisation ou reconversion?" *L'Actualité économique* 60: 471–94.

Lamonde, P. with M. Polèse, J. Lamoureux, and M. Tessier. 1988. *La transformation de l'économie montréalaise, 1971–1986: cadre pour une problématique de transport.* Rapports de recherches, no. 11. Montréal: INRS – Urbanisation.

Lamotte, A. 1985. *Les autres Québécoises: Etude sur les femmes immigrées et leur intégration au marché du travail québécois.* 2nd edition. Montréal: Gouvernement du Québec, ministère des Communautés culturelles et de l'Immigration.

Le Bourdais, C., P.J. Hamel, and P. Bernard. 1987. "Le travail et l'ouvrage: charge et partage des tâches domestiques chez les couples québécois." *Sociologie et sociétés* 19: 37–55.

Lewis, J. 1984. "The role of female employment in the industrial restructuring and regional development of the post-war United Kingdom." *Antipode* 16: 47–60.

Lipietz, A. 1980. "Inter-regional polarisation and the tertiarisation of society." *Papers of the Regional Science Association* 44: 3–17.

Maroney, H. 1983. "Feminism at work." *New Left Review* 141: 51–71.

Massey, D. 1984a. *Spatial Divisions of Labour: Social Structures and the Geography of Production.* London: Macmillan.

– 1984b. Foreword to *Geography and Gender: An Introduction to Feminist Geography,* Women and Geography Study Group of the IBG. 11–13. London: Hutchinson.

Massey D. and R. Meegan. 1982. *The Anatomy of Job Loss: The How, When and Why of Employment Decline.* London and New York: Methuen.

Menzies, H. 1982. *Women and the Chip.* Montreal: Institute for Research on Public Policy.

– 1985. "Une approche féministe des nouvelles technologies." In *Apprivoiser le changement: actes du colloque CEQ sur les nouvelles technologies, la division du travail, la formation et l'emploi,* 37–43. Montréal: Centrale d'Enseignement du Québec.

Mitter, S. 1988. "Flexible employment and poverty in the North: dimensions of race and gender." In *Women and the Economy: Conference Papers,* 19–33. London: Women's Studies Unit, Polytechnic of North London.

Molle, W. 1983. "Technological change and regional development in Europe." *Papers of the Regional Science Association* 52: 23–38.

Morokvasic, M., A. Phizacklea, and H. Rudolph. 1986. "Small firms and minority groups: contradictory trends in the French, German and British clothing industries." *International Sociology* 1: 397–419.

Mummé, C.L. 1983. "La renaissance du travail à domicile dans les économies dévelopées." *Sociologie du travail* 3: 313–35.

Myles, J. 1988. "The expanding middle: some Canadian evidence on the deskilling debate." *Canadian Journal of Sociology and Anthropology/Revue canadienne de sociologie et d'anthropologie* 25: 335–64.

Myles, J., G. Picot, and T. Wannell. 1988. "The changing wage distribution of jobs, 1981–1986" ["La répartition salariale des emplois: variations de 1981–1986"]. In *The Labour Force/La population active* (Oct.), 85–129. Statistics Canada, catalogue no. 71-001.

Nelson, K. 1986. "Labor demand, labor supply and the suburbanization of low-wage office work." In *Production, Work and Territory*, A.J. Scott and M. Storper, eds, 149–71. Boston: Allen and Unwin.

Newton, K. 1984. "Simple analytics of the employment impact of technological change." *Prometheus* 2: 233–45.

Noyelle, T.J. and T.M. Stanback. 1984. *The Economic Transformation of American Cities*. Totowa, NJ: Rowman and Allenheld.

Polèse, M. 1988. *Les activites de bureau à Montréal: structure, évolution et perspectives d'avenir*. Dossier Montréal 1. Montréal: Ville de Montréal and INRS – Urbanisation.

Polèse, M. and P. Lamonde. 1985. "La déplacement des activités économiques à l'intérieur de la région métropolitaine de Montréal: synthèse des résultats." Paper presented to the Annual Meeting of the Canadian Regional Science Association, 29–30 May, Montreal.

Power, M. 1988. "Women, the state and family in the U.S.: Reagenomics and the experience of women." In *Women and Recession*, J. Rubery, ed, 140–2. London: Routledge and Kegan Paul.

Pratt, G. and S. Hanson. 1988. "Gender, class, and space." *Environment and Planning D: Society and Space* 6. 1: 15–35.

Raffestin C. and M. Bresso. 1979. *Travail, espace, pouvoir*. Lausanne: Editions de l'Age d'Homme.

Rose, D. 1987. "Un aperçu féministe sur la restructuration de l'emploi et sur la gentrification: le cas de Montréal." *Cahiers de géographie du Québec* 31. 83: 205–24. English version, 1989. "A feminist perspective on employment restructuring and gentrification: the case of Montreal." In *The Power of Geography: How Territory Shapes Social Life*, J. Wolch and M. Dear, eds, 118–38.

Rose, D. and N. Chicoine. 1991. "Access to school daycare services: class, family, ethnicity and space in Montreal's old and new 'inner city'." *Geoforum* 22. 2: 185–201.

Rose, D. and P. Villeneuve. 1991. "Women workers and the inner city: some social implications of labour force restructuring in Montreal, 1971–1981." In *Life Spaces: Gender, Household, Employment*, C. Andrew and B.M. Milroy, eds, 35–64. Vancouver: University of British Columbia Press.

– 1993. "Work, labour markets and households in transition." In *The Changing Social Geography of Canadian Cities*, L.S. Bourne and D.F. Ley, eds, 153–74. Montreal: McGill–Queen's University Press.

Rubery, J. 1988. Introduction to *Women and Recession*, J. Rubery, ed, 3–14. London: Routledge and Kegan Paul.

Sassen, S. 1988. *The Mobility of Labour and Capital: A Study in International Investment and Labour Flow*. Cambridge: Cambridge University Press.

Sassen-Koob, S. 1984. "The new labor demand in global cities." In *Cities in Transformation: Class, Capital and the State*, Urban Affairs Annual Reviews, M.P. Smith, ed, vol. 26, 139–71. Beverley Hills: Sage.

– 1986. "New York City: economic restructuring and immigration." *Development and Change* 17: 85–119.

Sayer, A. 1984. *Method in Social Science: A Realist Approach*. London: Hutchinson.

Scott, A.J. 1982. "Production system dynamics and metropolitan development." *Annals of the Association of American Geographers* 72: 185–200.

Séguin, A.-M. and P. Villeneuve. 1987. "Du rapport hommes-femmes au centre de la Haute-Ville de Québec." *Cahiers de géographie du Québec* 31: 189–204.

Simmie, J.S. 1983. "Beyond the industrial city?" *Journal of the American Planning Association* 49. 1: 59–76.

Smith, J. 1984. "The paradox of women's poverty: wage-earning women and economic transformation." *Signs: Journal of Women in Culture and Society* 10: 291–310.

Smith, M.P. and J. Feagin. 1987. Introduction to *The Capitalist City*, M.P. and J. Feagin, eds, 3–21. New York: Blackwell.

Smith, N. 1986. "Gentrification, the frontier and the restructuring of urban space." In *Gentrification of the City*, N. Smith and P. Williams, eds, 15–34. Boston: Allen and Unwin.

– 1987. "Of yuppies and housing: gentrification, social restructuring and the urban dream." *Environment and Planning D: Society and Space* 5: 152–72.

Storper, M. 1981. "Toward a structural theory of industrial location." In *Industrial Location and Regional Systems*, J. Rees, G.J.D. Hewings, and H.A. Stafford, eds, 17–40. London: Croom Helm.

Toft-Jensen, H., P.A. Hansen, and G. Serin. 1983. "Capitalist technology and changes in the labour process." In *Technological Change and Regional Development*, A. Gillespie, ed, 89–103. London: Pion.

Villeneuve, P. 1989. "Gender, employment and territory in metropolitan environments." In *Labour, Environment and Industrial Change*, G.J.R. Linge and G.A. van der Knaap, eds, 67–86. London: Routledge.

Villeneuve, P. and D. Rose. 1986. "Force de travail et redéploiement industriel dans la région de Québec, 1971–1981." *Canadian Journal of Regional Science/Revue canadienne de science régionale* 9: 183–205.

– 1988. "Gender and the separation of employment from home in metropolitan Montreal, 1971–1981." *Urban Geography* 9: 155–79.

Villeneuve, P. and G. Viaud. 1987. "Asymétrie occupationnelle et localisation résidentielle des familles à double revenu à Montréal." *Recherches sociographiques* 28: 371–91.

8 Bargaining and Balancing: Women's Waged Work as an Adjustment Strategy in U.S. Households

SUSAN CHRISTOPHERSON

With the emergence of more fully internationalized product and financial markets that began in the early 1970s, industrialized countries have undergone a transition to a new regime of capital accumulation. Although the political-economic processes driving the transition to new production regimes oriented around production for international rather than domestic markets are, in some senses, supranational, they are also producing "new capitalisms" in differently situated national economies.[1] The resultant differences in national trajectories have in turn created different conditions for adjustment at the household and individual level.

In the United States structural transformation has been associated with an invalidation of the Keynesian productivity bargains that characterized the previous period of mass production oriented toward national markets. This transformation has brought about a redistribution of power, income, time, and knowledge among and within societies (Harvey 1989; Lash and Urry 1987). The literature examining household strategies in response to the economic conditions emerging from sectoral adjustment (e.g., worker lay offs or increasing work hours) or from state-imposed adjustment policies takes quite different forms. In the literature on household strategies in developing countries, state structural adjustment policies are at the center of the analysis because of their association with the debt crisis (which is a manifestation of the broader transformation of the world capitalist economy). The withdrawal of already minimal state services combined with high levels of unemployment lends credence to the use of the term "survival strategies" to describe household responses. In industrialized countries the picture is less grim, but the pressures on households to adjust to the withdrawal of state securities, to economic privation, or to declining standards of living are still present and can be accounted for within the same basic theoretical framework.

In industrialized countries in general, and in the United States in particular, there has been a tendency to favour resource generation strategies over better resource management strategies as adjustment or coping mechanisms. Some methods used to generate resources include: 1) increasing the supply of waged labour to the economy; 2) changing the assets/liabilities position of the household (increasing debt); 3) increasing the flow of income transfers; 4) tax reduction or evasion; 5) use of ethnic or religious community resources; or 6) participation in the informal economy.[2]

In studies of both industrialized and developing countries, feminist analysis of household responses has carried the interpretation of these processes a step further, examining how, within households, the burden of adjustment to more stringent economic conditions is distributed between men and women (Beneria and Roldan 1987). In what follows I examine evidence of structural transformation and adjustment in the United States using material from a study of nonmetropolitan communities.[3] In the United States regional and place-to-place variation in adjustment conditions and strategies is a much more important aspect of the adjustment story than it is in some other advanced industrial economies, particularly those in continental Europe, and reflects the role that spatial differentiation has played in the national adjustment process (Christopherson 1993). The intensified spatial differentiation resulting from both public and private investment decisions is visibly reflected in central city deterioration but is also manifested in intensified poverty and job redistribution in the nonmetropolitan areas of the country (Fitchen 1991; Gorham 1991).

The other distinctive feature of the U.S. adjustment path is the extent of differentiation in working conditions, including working hours and employment contracts. Employers in the United States have more flexibility in how they employ workers because they are less constrained than their industrialized country counterparts by public and private regulations governing working hours, lay offs, and employment-related costs such as job training. As a consequence employment conditions in the U.S. are more demand-driven and highly sensitive to output. This sensitivity was increased in the 1980s by changes in financial market regulation and the consequent pressure placed on firms by the high cost of capital. The result has been highly differentiated employment contracts; this differentiation is reflected in the increase in the proportion of the work force categorized as self-employed or temporary workers from 8 per cent in 1980 to 10 per cent in 1992 and in an increase in the proportion of the part-time work force that works as such involuntarily – that is, is under-employed.

These spatial and labour-force trends suggest that the transformation of contemporary capitalism in the U.S. cannot be captured by over-arching images of either growth or decline. Its signal quality is that of differentation, among demographic groups, within ethnic and racial groups, among segments of labour, and across communities. This fragmented pattern makes it difficult to see the national picture from the perspective of local case studies or to interpret what is happening locally on the basis of national trends. Two trends, though, are critical to household strategies across the U.S.: increased waged working hours, and increased waged work for women. These national tenden-cies provide a common ground for interpreting locally differentiated outcomes. So, before examining how the U.S. adjustment path is worked out in some particular nonmetropolitan locales, I will examine these national adjustment strategies.

ADJUSTMENT AT THE NATIONAL LEVEL: THE REDISTRIBUTION OF WORK

In the United States structural adjustment has involved a transition to an economy that, for several reasons, is dominated by employment in the services sector. First, the role of the United States, and particularly of certain United States regions in the international division of labour, has shifted to emphasize headquarters functions and business services. Secondly, the United States has lost massive numbers of manufacturing jobs as employers have pursued labour-cost reduction strategies to keep corporate profits up: 2.3 million manufacturing jobs disappeared during the 1981–2 recession, and almost another million were lost in the extended recession of the early 1990s. Thirdly, as female employ-ment has increased in reponse to the first two trends, there has been a trend to transfer services once provided in the home and in the public sector to the market (Thurow 1989).

Given the country's history of responding to economic difficulty by increasing income generation, it is not surprising that its most promi-nent household adjustment strategies since the 1970s have taken the form of increased work both in the household and for the individual. Since 1970, when the average workweek of the employed was 39 hours, working time has increased substantially. The average U.S. worker puts in approximately 164 more hours of paid labour per year now than in

1970, which is the equivalent of an additional month of work (Schor 1991).

A change in the distribution of work within the household and in women's work as it went from domestic to waged has also been a key feature of the transformation. Within the labour force there have been significant shifts in the sectoral distribution of women workers. In particular women have been affected as much or more than men by the loss of manufacturing jobs. Women were disproportionately employed in sectors such as apparel, textiles, printing, and consumer electronics, which have lost the highest proportion of jobs as production has relocated internationally. In contrast with male displaced workers, women have experienced longer spells of unemployment and greater losses of income as a consequence of plant closures and lay offs (Howland 1988). In addition employed women have had higher paying and more stable jobs in those communities where male employment has been in unionized, higher wage industries. With declining male employment, women's jobs in support or ancillary services associated with these sectors have also declined. Within service sectors women's jobs became increasingly bifurcated during the 1980s, with growth in female employment for high-skilled jobs in such fields as sales, accounting, and the professions, as well as in low-skilled service jobs.

The movement of women (especially those with children) into the United States work force along with the nationwide increase in working hours since the 1970s has meant that average household income levels have continued to increase for two-earner families. The dependence on multiple wage earners to maintain household income and to provide some income security has, however, widened the gap between those households in which there is more than one wage earner and those in which there is not. In the early 1990s this differentiation may be increasing because, for the first time since the 1960s, the flow of women into the job market has begun to fall (Uchitelle 1992).

The significance of this transformation is reflected in national and individual employment expectations. Women in the United States now expect and are expected to work whether or not they have young children; this stands in strong contrast to patterns of female labour-force participation in some industrialized countries, particularly in Japan and Germany, which have low rates of participation especially during the child-rearing period. This expectation exists in the U.S. despite the almost complete absence of state-supported social welfare provisions that are linked to the high rates of female labour-force participation in some industrialized countries, such as Sweden (Esping-Andersen 1990).

As was indicated earlier, one of the central elements defining the United States adjustment path is spatial differentiation in economic conditions, despite the sectoral convergence around service sector employment. Thus overall national employment and adjustment patterns conceal significant variations from place to place. To illustrate how local variations are constructed by sectoral adjustment in local context, the cases of nonmetropolitan areas and some specific nonmetropolitan localities will be examined.[4]

DIFFERENTIATED SPATIAL OUTCOMES: THE CASE OF NONMETROPOLITAN AREAS

Local economies in nonmetropolitan areas in the United States have undergone dramatic changes since the mid-1960s, becoming more integrated into national and regional economies as a result of infrastructural investment, technological advances in transportation and communication, and the dispersal of manufacturing and population. Industries historically dominant in nonmetropolitan economies – those based on agriculture, energy, and natural resources – have become increasingly sensitive to fluctuations in international and national markets.

The transformation of the nonmetropolitan economy has been widely perceived in terms of crises in particular sectors, such as agriculture and mining. At a global level many of the processes restructuring these industries are similar to those affecting urban-based manufacturing: the replacement of labour-intensive with capital-intensive production processes; the expansion of international sources for materials; and the internationalization of distribution and finance. What distinguishes the crisis of nonmetropolitan economies is the particular vulnerability of local communities to economic dislocation. Dependent upon employment in a single industry and frequently on a single firm, these economies have no escape from the processes of dislocation.

The effects of this transformation on local communities cannot simply be attributed to global dynamics. Responses to the globalization of production have been sectorally and firm specific. In addition communities have different political histories and resources. Some firms have substituted capital for labour (e.g., mining); others have changed the product mix (e.g., agriculture) or shifted the locus of production (e.g., forestry); all have increased subcontracted production (Young and Newton 1980). Since each industry initially had distinctive work patterns and cyclical sensitivities, communities have experienced this transformation in a variety of ways. For example those communities whose economies were traditionally based on waged labour had different sets of skills and coping mechanisms than had rural economies based on family farming. In this regard mining communities in the western United States stand out from other resource-based industrial communities in several respects. Their work forces are frequently unionized and have been paid relatively high wages. Also these economies typically have service sectors closely associated with the industrial base. Health-care services, for example, are more extensive in these communities than in other resource-based communities of similar size. They are not characterized by family labour but by male waged labour, and they have their own forms of adaptation to cyclical downturns, such as temporary outmigration (Bradbury 1989; Lund 1988). To put these general patterns in perspective, I will next examine some particular cases of local economic change and structural adjustment.

Cases of Adjustment in Arizona Nonmetropolitan Communities

Arizona, the sixth largest state in the United States in terms of territory, has a highly clustered nonmetropolitan population, which is largely a function of its desert climate and the association of settlement with natural resource extraction. Arizona has experienced growth in its population along with certain other states because of in-migration to retirement and recreation communities, affecting some counties of the state more than others – particularly those adjacent to the major metropolitan areas of Phoenix and Tucson (Mulligan and Reeves 1986). Although growth slowed in the 1980s these areas were not as adversely affected by the boom/bust cycle of manufacturing-induced nonmetropolitan development as were other nonmetropolitan areas in the country because of their greater dependence on stable income sources, such as transfer payments to retired and Native American populations. Although the Arizona economy continues to include primary sector export industries, these play a decreasing role relative to locally oriented sectors, particularly retail trade, tourism, and construction (Silvers and Shelnutt 1987; Valley National Bank 1985). This is especially true in nonmetropolitan Arizona. And while, in comparison with some northern states, manufacturing employment has grown in Arizona during the past twenty years, that growth is almost exclusively attributable to aircraft and missile production and, thus, dependent on defense expenditures (Silvers and Shelnutt 1987). Defense reductions can be expected to have a significant impact on this employment base.

Research on economic change in Arizona nonmetropolitan communities in the 1970s and 1980s details two intersecting patterns.[5] First, as towns in the counties adjacent to nonmetropolitan areas expanded consumer service functions, their economies became disarticulated from those of the counties, many of which still have a significant industrial base in primary sector industries (Arizona, Department of Commerce 1986). As these industries were restructured, however, they employed fewer people. Secondly, although all nonmetropolitan communities show growth in service employment between 1975 and 1986, there are some significant differences in the types of service jobs that exist from one community to another. As service employment grew in proportion to total employment, differences in service economies emerged in terms of four factors not captured by broad sectoral categories. The first two of these, which describe the organization of production, are establishment size and the relationship of service functions to other sectors of the local economy; the third and fourth describe the organization of work: the distribution of work time, and the employment relation – specifically, the degree to which work is carried out by hourly, by self-employed, or by quasi-self-employed workers. These factors describe differences among local nonmetropolitan economies, all of which are gaining in service employment.

Four distinct types of local nonmetropolitan service economies appear to be emerging in Arizona. The first is a traditional market economy still prevalent in some counties that are not adjacent to metropolitan areas. The services of these communities retain strong links to and dependence upon primary sector employment in the county, and their economies are oriented toward local consumption. The second type is the service economy characteristic of mining towns that are notable for a highly developed social-service sector and service inputs into the primary sector, for example in machine repair. The third type is the government-oriented service economy, which employs large numbers of administrative staff in full-time jobs and at wages relatively closer to those of urban centres. The fourth type is the "new" service economy associated with population growth related to retirement or recreation activities; it is characterized by more part-time work, by self-employment or quasi-self-employment (work for commission), and by growth in small competitive service firms. Growth in these part-time jobs is strongly associated with rapid increases in female labour-force participation. In the majority of the communities in the sample from the Arizona community data base, more than 50 per cent of the employed women work full-time, although almost all are subject to seasonal lay offs or hours reductions. The mean percentage of women working full-time for this sample is 70 per cent, but the mean of the five communities with the lowest rate of full-time workers among the female labour force falls to 44 per cent.

A regression analysis that explains differences in the percentage of women working full-time throughout the full set of communities identifies four variables that account for 58 per cent of the variance: adjacency to a major metropolitan area; population size; the female labour-force participation rate; and the per cent change between 1970 and 1980 in women in the labour force at the county level. Interestingly a higher total number of full-time workers in a community is inversely related to the percentage of women employed. In addition an increase between 1970 and 1980 at the county level in the percentage of women employed was inversely related to the percentage of women working full-time in the nonmetropolitan communities located in those counties. This finding lends support to the notion that the new jobs created for women have been predominantly part-time jobs.

Probably the most important finding regarding women's employment in these communities is that the decline in primary sector employment has affected women's employment as well as that of men. Although counties and communities dominated by primary sector activity, particularly mining, have low female labour-force participation rates, in the past those women who did work were likely to hold full-time jobs. In the communities where primary sector activity has declined, there has been a significant drop in full-time employment for women. State data, corroborated by the household interviews, also indicates that the decline of primary sector employment is associated with a decline in public sector employment or with a decrease in working hours for public employees. Since the public sector is a major provider of female jobs, this loss also has adversely affected women's employment opportunities.

A second important finding is that much of the growth in employment for nonmetropolitan women in a high growth state such as Arizona is part-time employment. In those towns dominated by primary sector employment, the relatively low female labour-force participation rates began to rise in the 1980s, but new jobs were more likely to be part- than full-time.

Three Patterns of Adjustment

The three communities chosen as case studies represent different patterns of adjustment to declining primary sector employment and increasing employment in services.[6] Globe-Miami is an old mining town only recently affected by significant permanent lay offs in the mining work force. Its employment profile, which includes the service sector, still reflects dependence on primary sector employment. The second community, Bisbee, lost mining employment in the early 1970s, and its recovery is attributable to a large and growing federal government sector. The third community, Payson, a new service economy, has grown the most rapidly of the three communities in population and employment; but its employment profile is that of a consumer service economy, with a larger proportion of part-time jobs, low-wage jobs, and more self-employment or quasi-self-employment, such as sales commission jobs.

The employment patterns emerging in nonmetropolitan Arizona in the 1980s reflect in microcosm the new sources of differentiation in the United States work force described in my introduction. In the communities formerly dominated by mining industry employment the proportion of full-time jobs has declined. And, because these economies have been structured around full-time jobs in mining and the public sector, there has been less leeway for adaptation through part-time work for women; thus unemployment rates are higher. In those communities where service employment has always predominated or in new service economies emphasizing retirement and recreation, employment expansion has tended to take the form of part-time jobs. The implications of these different economic paths become clearer if we look at three communities and at the nature of service sector employment in more detail.

A functional classification of service industries allows for a more detailed examination of service activity in these three communities than that based upon the traditional employment by sector. In particular we can disaggregate the relative importance of consumer services, social services, producer services, and distributive services. This type of analysis suggests two areas of differentiation that go considerably beyond sectoral distinctions. First, while the distribution of firms by sector is relatively similar across all three communities (by far the largest number of firms are in consumer services), Globe-Miami and Bisbee are notable for having a higher portion of employment in large firms or organizations. This pattern extends beyond the primary sector, where it could be expected, to the service sectors of these two communities. Only in producer and consumer services does Payson have marginally larger firms, and these are at the very low end of the firm size spectrum. Across the three communities there appears to be a tendency for employment decline in continuously existing firms, particularly in male-employing primary and secondary sector firms; as a consequence, larger firms are getting smaller, and smaller firms are growing.

These communities differ in their employment-per-firm profile as much or more than in their sectoral employment distribution. This difference is significant insofar as employment in small firms is generally associated with more part-time and fewer full-time jobs, lower wages and benefits, and higher turnover (United States Small Business Administration 1985, 1986, 1987). The firm size implications become clearer if we look at the three communities in more detail.

Payson, northeast of Phoenix, is the smallest of the communities, with a population of approximately 7,000. In the early 1970s a significant portion of Payson's employment was concentrated in mining and quarrying (27 per cent) and manufacturing, principally in a lumber mill (11 percent). Cattle ranching also played a role in the economy of the county surrounding the town. Retail trade employed 13 per cent of the labour force, and services, 10 per cent (Arizona Department of Economic Planning and Development 1972). By 1986 agriculture, mining, and manufacturing together employed only 7 per cent of the Payson labour force, while retail trade employed 38 per cent and services, 29 per cent. The transformation of Payson is related to its proximity to Phoenix (94 miles away) and nearby national forest land. While not a tourist destination, it is a "drive through" (i.e., a stop on the way) to other destinations. It has become a summer home location for Phoenix residents as well as a low-income retirement community (Arizona Department of Commerce 1986). Much of the growth in this town is, in fact, attributable to in-migration. It is a town of newcomers rather than long-time residents.

Payson exemplifies the high-growth, service-oriented local economy. Data from the Arizona community data base indicate quite dramatic growth in overall employment: the number of people employed in Payson grew by 152 per cent from 1976 to 1985 and, during this nine-year period, the female labour-force participation rate grew from 43 to 49 per cent, while the percentage of full-time jobs declined by 7 per cent and part-time jobs grew by 6 per cent. Retail employment has consistently accounted for the majority of jobs in the community, increasing over these nine years from 34 to 36 per cent. Within the retail sector, however, male full-time employment has declined (from 15 to 11 per cent) while female part-time employment has increased (from 4 to 8 per cent). In general terms services accounted for 30 per cent of the employed work force in 1976 and 29 per cent in 1985.

Payson is notable among the three communities for having a higher portion of its employed work force in producer services. This is somewhat misleading, however, since employment in this sector was almost exclusively located in real estate firms. Thus Payson, while growing rapidly, is an economy dominated by small firms and relatively low-paying and less-reliable service jobs. Related to the small-firm, service-based character of the economy are a sizeable number of jobs in which the worker's income is substantially dependent on individual effort whether or not he or she is self-employed. Examples include real estate or insurance sales and construction jobs.

Globe-Miami, also in Gila County, contrasts sharply with Payson in several respects. Copper mining was responsible for the establishment of Globe-Miami at the turn of the century; until the early 1980s, the local economy was dominated by employment in that industry. Approximately 40 per cent of the labour force was employed in the mining or agricultural sectors in 1978; this figure had decreased to 32 per cent in 1986. Employment declined sharply with mining lay offs that reduced the number employed from 6,686 in 1978 to 6,112 in 1986 (a 9 per cent decline). There has also been a decline in the number of full-time jobs and an increase in part-time employment. Female labour-force participation rates are lower than those of the other two case study communities but have increased from 34 per cent in 1978 (27 per cent full-time and 7 per cent part-time) to 42 per cent in 1986 (30 per cent full-time and 12 per cent part-time). Most of this growth was concentrated in service sector jobs. The Globe-Miami economy is now more sectorally diversified, with a 10 per cent increase in service employment (from 21 per cent in 1981 to 31 per cent in 1986) and marginal increases in public administration and retail trade employment.

Bisbee, the third case study town, was also a mining town at its inception. Copper mining and smelting employed approximately 17 per cent of the labour force until Phelps Dodge shut down the Lavender Pit Copper mine in 1975. Mining and associated manufacturing employed only 8 per cent of the labour force by 1986 (Arizona Department of Commerce 1986). Bisbee's location on the United States–Mexico border and a large adjacent military center (Fort Huachuca), however, have combined with an emerging tourist industry to make Bisbee's economy more truly diversified, both with respect to the industrial sectors represented and the types of jobs available.

As in the other two towns, female labour-force participation rates have risen in Bisbee and much of the employment growth has been in services. Rates of female labour-force participation for Bisbee were 48 per cent in 1981 and 51 per cent in 1986. While the portion of full-time female workers remained the same (37 per cent), the percentage of part-time workers increased from 11 to 14. The most notable feature of this local economy is its continued dependence on public sector employment. In 1986, public administration employed 36 per cent of the local labour force, up from 31 per cent in 1972. While overall male full-time jobs declined from 44 per cent in 1981 to 39 per cent in 1986, in the public sector they increased from 12 per cent in 1981 to 15 per cent in 1986. Male full-time employment decreased in services (from 10 to 8 per cent), construction (from 4 to 2 per cent), as well as mining and agriculture (from 4 to 2 per cent).

So, like the country as a whole, all three of the case study communities have demonstrated substantial growth in their service sectors. There are, however, some special characteristics of their service employment patterns. First, the high-growth business service sector that is fueling metropolitan growth is negligible in all of these communities. The few small business service firms are confined to the very standardized low-wage services, such as building maintenance, protective services, and photo duplication. In general growth in the first half of the 1980s in these communities was concentrated in consumer services and in public sector services related to population size, such as schools. Major employment growth was, as typical of the country, in retail establishments and in health services, and these establishments have tended to hire a high proportion of female part-time workers. Although there was an apparent relationship between the expansion of the service sector and increasing female employment in these communities, this relationship gives us only a partial picture of how households and individuals are responding to altered economic opportunities. To interpret the implications of economic adjustment we need a broader conception of the alternatives available to households. These are explored in the household interviews conducted in conjunction with this research.

HOUSEHOLD ADJUSTMENT STRATEGIES

Given the changing economic orientation of the three communities and their emerging patterns of labour demand, how have households responded? Information gathered from household interviews in the three case study communities supports the view that households have adapted to economic difficulty by increasing female waged labour. This strategy is used in combination with other resource generating strategies, such as long distance job searches and unreported or "off-the-books" employment. In some respects female employment appears to play a role similar to that of farming households for migrant labour: a stable "fall-back" for more unpredictable but more highly remunerated work.

The jobs worked by female respondents in our sample averaged 43.5 hours per week, while males worked an average of 47.4 hours per week. Females contributed 42 per cent of total household income in comparison with 58 per cent for males. The disparity between hours worked and income contribution is a consequence of lower female incomes. Female respondents had an average income of $14,016 and males, $22,217. Among the three towns there were considerable differences in the distribution of household income. Globe-Miami, the traditional mining community, had the highest incomes and the narrowest distribution. Payson, the high-growth new service economy, had the widest income distribution among the households in the sample. These relative income profiles are comparable with state and county data on household income levels in the three case study towns.

As has already been noted, women's substantial contribution to household income is based on a high number of average hours worked per week (for those women who generally work full-time). If women's wages were equal to those of men, they would be contributing half of total household income. As it is they work, on average, almost as many hours as men but earn substantially less. Given that women continue to do more work in the home, their workweek is substantially longer than that of men.

Both the pattern of employment among the three communities and comments by the respondents suggest that as nonmetropolitan communities move from male-employing primary sector industries to female-employing service industries, the traditional interpretation of women's part-time work as voluntary may be less valid. With the expansion of services, women are even more segregated by industry and occupation than was previously the case. This occupational segregation, which was evident in our sample, is more pronounced in nonmetropolitan areas than in metropolitan areas. And, since part-time work is concentrated in those few sectors with a predominantly female labour force, viable full-time work options at a sufficient wage (enough to pay for child care, for example) are not available. Consequently part-time work becomes the only viable option and not a choice in any accepted sense.

The findings from our interviews support the view that demand as well as supply affects female employment patterns and that the concept of voluntary part-time work is significantly open to question. The employment of women in part-time jobs does not appear to be explained solely as a matter of choice. Despite the similarity of the female respondents with respect to age, education, and family characteristics, their employment patterns differ from one community to the next. In Bisbee, the community in which well-paid public sector full-time work is available, more women hold full-time jobs. In contrast in Payson, where many of the new jobs being created are part-time jobs, more of the respondents hold more than one part-time job. The questions raised by the findings in this study regarding voluntary part-time work are supported by other recent research, which suggests that women in the United States now place a premium on maintaining labour-force attachment throughout their lives and assume that they must combine work and family. In part this change is attributable to the fact that women's income is no longer supplementary to household income but is central (Paulson 1982; Seitchik 1991). The findings also support research on the nature of the part-time work force, which indicates that a smaller number of part-time workers would choose to continue to work part-time if adequately paid full-time work were available to them (Christopherson and Noyelle 1992).

Different Work or More Work?

Throughout Arizona growth in service employment has increased the demand for female labour but almost exclusively in the traditional female-dominated occupations of clerical and service work. That men's and women's occupational and industrial profiles are converging is attributable to a general narrowing of the occupational profile for both sexes rather than to the movement of women into occupations and industries in which they were previously underrepresented (Megdal 1986). Women now constitute a higher proportion of total employment in traditionally female occupations (sales and clerical work) than they did in the 1960s. This concentration in a few sectors and occupations is even more characteristic of female employment in the non-metropolitan areas of Arizona than it is in the metropolitan areas.

While nonmetropolitan women are increasingly employed in the same sectors and in similar occupations as metropolitan women, differences in their relative wages and in working hours have become more significant. Analysis of changes in women's employment nationwide indicates that the new sources of differentiation in the labour force are intraoccupational and intrasectoral (Gorham 1991). As more of the work force is employed in services, local labour-market conditions, firm size, the nature of the employment relation, and access to work hours will be relatively more important in creating differences among workers than will employment in a sector or occupation (Christopherson 1989).

These new sources of labour segmentation in nonmetropolitan economies are also demonstrated in the interview findings. Women respondents in relatively skilled, nontraditional occupations, such as newspaper editors, earn lower wages than they could command in metropolitan areas. The majority of women with full-time jobs work over 40 hours per week. Comments from women with either full-time or part-time jobs suggest that they anticipate having to look for a second job in order to "make ends meet." These findings, which are suggestive rather than definitive, are nonetheless supported by national data recently developed by the Economic Policy Institute, which indicate that women's hours of work in the United States have increased by 8.5 per cent over the past five years. The increased earnings of women are at least partially attributable to more waged work rather than to increased pay. In addition the choice may not be between a 20-hour-per-week job and a 40-hour-per-week job but between a part-time job, the hours of which are frequently extended, and a 50-hour-per-week, full-time job. Thus the new nonmetropolitan service economy may differ less in the kinds of jobs available than in the number of hours of work available.

CONCLUSION

It would be presumptuous to make any definitive statements based on the results of a study that only begins to delve into questions of nonmetropolitan economies as service economies. The results of this study, however, raise questions about both current interpretations of economic change in nonmetropolitan areas and the implications of emerging female employment patterns. These questions deserve broader, more systematic analysis.

We know that nonmetropolitan economies were always cyclically sensitive and periodically affected by the crises of restructuring and rationalization in primary sector industries. Local economies, work forces, and households developed coping strategies adapted to these periodic cyclical crises (Bradbury 1989). The information presented in this study suggests that as local economies become more dominated by service employment, one type of cyclicality, that based in the crises associated with primary-sector industrial production, is replaced by others. These new forms, such as the seasonal cycle of tourist employment, are significantly different from those that preceded them; they are, for example, even more difficult for the local work force to influence or to predict than were previous cycles. Therefore new patterns of adaptation at the level of the community and the household come into being.

A related conclusion concerns our understanding of the nature of the service sector, of its spatial distribution and employment patterns. Much of our understanding of the transformation to service-based economies at the local level is obscured by the measures we have used to analyse economic change. There are at least two identifiable problems that are beginning to attract more attention, the first of which is our reliance on simple sectoral measures to understand employment patterns. Though sectoral distinctions were never completely reliable for predicting employment relations, they are even less so in this era of complex interfirm and interindustry relationships. The second and closely related problem is our inadequate conceptualization of services and of how they are evolving and becoming differentiated as they expand within the United States economy. For the most part, our concept of services is still fixed in the simple notion – adequate in the 1950s when "services" referred primarily to consumer services ancillary to employment in manufacturing – that relates service growth to population growth. Only recently have there been attempts to take apart service activities and examine them in close detail; as for the two most generally accepted recent categorizations, one distinguishes producer services from consumer services, and the other distinguishes service activities according to rates of employment growth (Thurow and Waldstein 1988). A recent effort distinguishes tertiary or passive services from those growth-inducing services that are characterized by information and knowledge intensity (Appelbaum and Albin 1990). Others distinguish market and nonmarket services and break down consumer services according to the scale of their final personal or social, market. Until we have made progress in understanding the organization of service activities, it will be difficult to predict locational tendencies in different types of services.

The continuing expansion of service employment has made it more difficult to evaluate differential shifts in the employment profiles of metropolitan and nonmetropolitan areas. Within the boundaries of the existing categories, sectorally based differences appeared to have converged, particularly during the 1970s when metropolitan areas were losing the more sectorally diversified manufacturing jobs. It is becoming apparent, however, that this convergence cloaks other, increasingly important, forms of divergence based in the organization of production, in firm size, interfirm relations, and degree of specialization (routine or nonroutine activities). Spatial differentiation may express itself in terms of job quality, wages, and hours worked rather than in traditional terms of industry or occupation (Christopherson 1989).

Finally this study supports and further illuminates national studies, which suggest that while there is some evidence that United States women are moving into nontraditional and better-paid occupations, an increase in women's hours of work at least partly accounts for their (now eroding) income gains relative to men and for the relative stability of household income levels. Since this waged labour has costs (for example, in transportation and purchased services such as fast food), it is not clear whether this additional cash income translates into an improvement in living standards.

NOTES

Research for this paper was supported by a grant from the Ford Foundation and Aspen Institute through the Southwest Institute for Research on Women, University of Arizona, Tucson. I would like to thank Ellen Hansen and Wendy McFarren for research assistance. Linda Peake and Audrey Kobayashi generously provided comments.

1 These new capitalisms result from the way that persistent national rules governing firm behaviour and financial markets construct different national outcomes. See Christopherson 1993.

2 For a broader discussion of household strategies, see Cornia 1990.

3 The study addresses questions at several geographic and socioeconomic levels. Information on community economic bases and on the role of female labour within local economies is derived from an analysis of unique data based on nonmetropolitan communities developed by the Department of Geography and Regional Development at the University of Arizona. These data are derived from surveys administered to all employers in thirty-two nonmetropolitan Arizona communities between 1975 and 1986. The questionnaires obtained information on full- and part-time employment by gender and sector and on seasonal employment by sector. To look at change over time, this data set was supplemented by census data (1970 and 1980) on employment and population characteristics; by Bureau of Economic Analysis estimates of personal income; and by County Business Patterns data on the number of firms and workers for 1964, 1974, and 1984. Three communities were then selected for case studies. This selection was based on two criteria: 1) availability of detailed information on the local economy for at least two points in time during the survey period; and 2) representation of different types and phases of adjustment to primary sector industrial decline. Of the three localities selected, Payson is an economy that has expanded very recently and is highly concentrated in "new" service employment; Bisbee, once heavily dominated by mining employment, is diversified but heavily dependent on a strong government sector; Globe-Miami, a community dominated by the mining industry, has recently lost that source of employment. The analysis of economic change in these three communities concentrates on the employment characteristics of service sector growth in each community.

To answer questions related to household adjustment strategies and to the role of female employment in emerging service-based nonmetropolitan economies, 153 household interviews were conducted with married, employed women, equally distributed through the three case study communities. Since the interviews were intended to raise questions for future, more systematic research (and because of funding limitations), a "snowball" technique was used to identify interviewees. The interviews obtained data in three areas: household composition; sources of income; and origin and use of contributions by household members.

4 Another major reason for looking at the variation of restructuring processes across space is political, in the sense that it is at the local level that people mobilize politically and either resist or adapt. Our ability to interpret the possibility of each of these responses depends upon an interpretation of how pressures and capabilities are formed in some places as opposed to others.

5 Arizona has some distinctive demographic features that intersect with economic trends. Most notably it has nonmetropolitan counties with sizable minority populations – Navaho and Pueblo, as well as other Native American peoples and Mexican-Americans.

6 A shift-share analysis examining the relative growth or decline of sectors in the three communities in comparison with the state as a whole shows that Bisbee and Globe-Miami have sectors that grew more slowly than did these industries statewide in comparable years. In Bisbee all industrial sectors grew at a slower pace than in the state as a whole with the exception of public administration (+.33), which offset decline in other sectors. The total differential shift in relation to the state was −.22. Globe-Miami suffered even greater losses relative to state growth rates, particularly in finance (−1.06), construction (−.98), and manufacturing (−.77). The total differential shift of −.65 indicates the impact of the employment losses in Globe-Miami relative to the general growth in employment in the state as a whole and particularly in the metropolitan areas. In contrast with Bisbee and Globe-Miami, Payson much more closely resembles state employment growth patterns with a total differential shift of +.89.

REFERENCES

Appelbaum, E. and P. Albin. 1990. "Shifts in employment, occupational structure, and educational attainment." In *Skills, Wages, and Productivity in the Service Sector*, T. Noyelle, ed, 31–65. Boulder: Westview Press.

Arizona. Department of Commerce. 1986. *Arizona: 1986 Economic Profile.* Phoenix.

Beneria, Lourdes and Martha Roldan. 1987. *The Crossroads of Class and Gender: Industrial Homework, Subcontracting, and Household Dynamics in Mexico City.* Chicago: University of Chicago Press.

Bradbury, J. 1989. "Strategies in local communities to cope with industrial restructuring." In *Labour, Environment and Industrial Change*, G.J.R. Linge and G.A. van der Knapp, eds, 161–76. London: Croom Helm.

Christopherson, S. 1986. "Women in the Arizona economy: recent trends and future prospects." In *Women and the Arizona Economy*, J. Monk and

A. Schlegel, eds, Tucson: Southwest Institute for Research on Women, University of Arizona.

— 1989. "Flexibility in the U.S. service economy and the emerging spatial division of labour." *Transactions of the Institute of British Geographers* 14: 131–43.

— 1993. "Market rules and territorial outcomes: the case of the United States." *International Journal of Urban and Industrial Research* 17.2: 274–88.

Christopherson, S. and T. Noyelle. 1992. "The U.S. path toward flexibility and productivity." In *Regional Development and Contemporary Industrial Response*, H. Ernste and V. Meier, eds, 163–78. London: Bellhaven.

Cornia, G. 1987. *Adjustment with a Human Face*. Oxford: Clarendon Press.

Esping-Andersen, G. 1990. *Three Worlds of Welfare Capitalism*. Princeton: Princeton University Press.

Fitchen, J. 1991. *Endangered Spaces, Enduring Places: Change, Identity and Survival in Rural America*. Boulder: Westview Press.

Gorham, L. 1991. "The slowdown in nonmetropolitan development: the impact of economic forces and the effect on the distributiton of wages." In *Population Change and the Future of Rural America*, David Brown and Linda Swanson, eds, 142–53. Washington D.C.: Economic Research Service, U.S. Department of Agriculture.

Harvey, David. 1989. *The Condition of Postmodernity*. Oxford: Basil Blackwell.

Howland, M. 1988. *Plant Closings and Worker Displacement: The Regional Issues*. Kalamazoo: W.E. Upjohn Institute for Employment Research.

Lash, S. and J. Urry. 1987. *The End of Organized Capitalism*. Oxford: Oxford University Press.

Lund, M. 1988. Industrial decline and readjustment: the experience of Minnesota's iron range. Master's thesis, School of Industrial and Labor Relations, Cornell University.

Megdal, S. 1986. "Women in the Arizona economy: a profile." In *Women in the Arizona Economy*, J. Monk and A. Schlegel, eds, 64–98. Tucson: Southwest Institute for Research on Women, University of Arizona.

Mulligan, G. and Richard W. Reeves. 1986. "Employment data and the classification of urban settlements." *The Professional Geographer* 38. 4: 349–58.

Paulson, N. 1982. "Change in family income position: the effect of wife's labor force participation." *Sociological Focus* 15. 2: 77–91.

Schor, J. 1991. *The Overworked American: The Unexpected Decline of Leisure*. New York: Basic Books.

Seitchik, A. 1991. "When married men lose jobs: income replacement within the family." *Industrial and Labor Relations Review* 44. 4: 692–707.

Silvers, A. and J. Shelnutt. 1987. "Sectoral bases of economic growth in Arizona: 1967–1986." *Arizona Review* 35. 1: 19–35.

Thurow, L. 1989. "Regional transformation and the service activities." In *Deindustrialization and Regional Economic Transformation*, L. Rodwin and H. Sazanami, eds, 179–98. New York: Unwin Hyman.

Thurow, L. and B. Waldstein. 1988. *Services in the American economy.* Washington, D.C.: Economic Policy Institute.

United States. Small Business Administration. 1985, 1986, 1987. *Annual report*. Washington: GPO.

Uchitelle, L. 1992. "The Undercounted Unemployed." *New York Times*, 10 January, sec. C, pp. 1–2.

Valley National Bank of Arizona. 1985. *Arizona statistical review*, September. Forty-first annual edition.

Young, J. and J. Newton. 1980. *Capitalism and Human Obsolescence*. Montclair, NJ: Osmun and Co.

9 Gentrification, Work, and Gender Identity

LIZ BONDI

The term "gentrification" has gained widepread currency in recent years. It refers to a process whereby relatively affluent people become residents of less affluently populated urban localities, bringing about the renovation or rebuilding of the built environment and attracting new cultural facilities. The process is related to changes in both paid and unpaid work: it is associated both as cause and as effect with the growth of service sector employment and the emergence of a service class that is establishing its own niche within occupational and residential hierarchies (Thrift 1987). It also signals a shift, identified by Gershuny (1983), from the purchasing of final services to the purchasing of intermediate services and consumer durables, to which the consumer adds his or her own labour in the creation of an apparently distinctive home environment.

These changes in the character of work are themselves related to changes in the position of women within both the formal and the domestic economies. Consequently one aspect of gentrification that has received frequent mention is the role of women, particularly in relation to changes in household structure and in the gender composition of the labour force. For example Beauregard (1986, 43) draws attention to the importance of "a trend toward the postponement of marriage and childbearing," while Markusen (1981, 32) cites "the success of both gays and women in the professional and managerial classes in gaining access to 'decent-paying jobs.'" As Smith (1987, 156) notes, however, the association between women and gentrification "has remained a general affirmation with little documentation of actual trends." Moreover little attempt has been made to locate the role of women in gentrification within broader discussions of gender relations, although it is clearly assumed that changes in the position of women, both in the family and in the paid labour force, have been influential. This essay addresses this lacuna in the literature and explores the interrelationships between women, work, and place in the context of the gentrification of small localities within contemporary Western cities. In the first section, existing feminist research on the relationship between gender, work, and urban spatial structure is outlined. This research demonstrates how past changes in the spatial organization of work have influenced the roles of women and men and how, conversely, changes in the gender division of labour have contributed to the development of particular spatial arrangements. The second section examines the concept of gentrification and emphasizes how recent debates have highlighted, although not necessarily investigated, questions concerning the gender division of labour. Against this background, the third section sets out existing propositions regarding the role of women as paid and unpaid workers in the processes of gentrification. The consequences of such developments for gender relations remain little understood, and in the fourth section the concept of gender is discussed, with attention focused on the interplay between its objective and subjective facets. This leads finally to a reconsideration of the role of women in gentrification in terms of the relationship between household practices, housing choices, and patterns of work.

WOMEN, WORK, AND URBAN STRUCTURE

Analyses of urban development have often cited changes in the organization of production and employment as major causal factors. Until recently the gender dimension of such changes remained neglected, but this is now being rectified in studies by feminist geographers. For example Mackenzie and Rose (1983) and McDowell (1983) have argued that the separation of home and workplace, precipitated by the advent of factory production, became a key factor in the subsequent emergence of a distinctively female domestic realm. Domestic work, separated from production for exchange, became differentiated from other types of work by virtue of being both unpaid and undertaken by women. In this instance changes in gender divisions are interpreted as a consequence of changes in both the spatial organization of urban areas and the organization of production: the definition of men as breadwinners and women as homemakers spread through social groups as a result of a) the removal of productive activities from the home, and b) the spatial separation between residential areas and places of work.

Another example concerns the growth of suburbia. Davidoff, L'Esperance, and Newby (1976) demonstrate the close links between this residential form and the idealization of a particular family form, namely a nuclear family unit with the husband as paid worker and sole breadwinner and the wife as an unpaid and economically dependent, full-time mother and homemaker. Here it is argued that the aspiration to a particular social arrangement and gender division of labour was intrinsic to the desire for a particular residential form. Subsequently changes in the gender division of work in the formal economy, manifested most clearly by the reentry of married women into the labour market since the early 1960s (Beechey 1986), have provided a powerful factor in the locational decisions of certain kinds of employer. Thus Massey (1983) and Lewis (1984) assert that geographical aspects of contemporary industrial restructuring in Britain are strongly influenced by the characteristics of female labour pools created by regional differences in gender divisions of work. In particular they suggest that regions previously dominated by coal mining and heavy industry have now become attractive to secondary and tertiary industries because of the presence of a female work force unaccustomed to waged labour and with no tradition of labour organization.

At the urban scale, McDowell (1983, 67) advances a similar argument: "as the economy expanded [in postwar Britain], the pool of women on ... suburban estates proved an attractive and flexible source of labour for the light assembly industries that also began to decentralise and expand in the suburbs." At a more detailed level, Lewis and Foord (1984) and Tivers (1985) examine how the urban environment affects the daily lives of women, focusing on two British New Towns and a London suburb, respectively. Both studies show that conventional urban planning, particularly the strict separation of residential and nonresidential land uses, militates against attempts by women to combine raising a family with paid employment. As a consequence women with young children tend to be confined to a domestic existence in the suburbs. Furthermore women who are in paid employment are likely to work much more locally (and for shorter hours) than are men. Consequently there is a strong association between women and suburbs, on the one hand, and men and city centres, on the other (Saegert 1981).

Wekerle (1984) argues to the contrary, however, drawing on North American evidence to claim that "a woman's place is in the city." For example Freeman (1981) reports that women form a greater percentage of the residents of inner-urban than of suburban or nonmetropolitan areas. In part this trend reflects the preponderance of poor households (usually Black) in inner cities (Holcomb 1984), a group including most households headed by women and many elderly women living alone; it is not, therefore, the product of choice. Nevertheless there is also some evidence (again from North America) that women actively prefer central to suburban locations (Saegert 1981). Thus, whether rich or poor, whether by choice or necessity, women appear to be breaking their traditional association with the suburbs.

These studies hint at the possibility that gentrification might be causally related to contemporary changes in the socioeconomic position of women in society: Saegert's evidence suggests that inner-city living might express the residential choices of affluent women (especially those in professional employment), while evidence regarding the feminization of poverty suggests that the economic vulnerability of other women may be contributing to the opportunities for gentrification. Alongside the position of women in the formal economy are changes in the domestic economy and home life, especially the increase in marital breakdown. These trends have prompted Markusen (1981, 32) to assert that "gentrification is in a large part a result of the breakdown of the patriarchal household." However, as she acknowledges, research to support this speculation is lacking. Certainly, despite growing interest in the relationship between geographical patterns and gender divisions, feminist discussions have, as yet, given little consideration to the issue of gentrification. Attention must therefore turn to the voluminous literature on gentrification.

GENTRIFICATION, HUMAN AGENCY, AND WORK

Despite its familiarity and apparent clarity, the term "gentrification" has been the subject of considerable debate. It has been described as a "chaotic conception" (Rose 1984), and three different elements of ambiguity can be identified within attempts to define it. First, there are disagreements about whether gentrification is, by definition, an inner-city phenomenon. Williams (1984) argues that the term should not be restricted to a particular location, but Munt (1987) argues on both empirical and conceptual grounds that inner-city location must be regarded as a defining feature of gentrification. The evaluative connotations of related terms have also provoked disagreements: Palen and London (1984) and Ley (1981, 1986) refer to "inner-city revitalization," while Lauria and Knopp (1985) refer to "urban renaissance." Smith (1982, 139) rejects these terms on the grounds that they "suggest that the neighborhoods involved were somehow de-vitalized or culturally moribund. While this is sometimes the case, it is often true that

very vital working-class communities are de-vitalized through gentrification." These semantic disagreements are indicative of a failure to
identify intrinsic elements or necessary relations, from which the contingent aspects of any empirical example can be distinguished. The
second source of ambiguity in the concept concerns the relationship
between changes in the built environment (which result in gentrified
residences), and social changes, including changes in the nature of
employment (which result in the presence of "gentrifiers"). Analyses
typically subsume both kinds of processes under one term without
distinguishing between cause and effect. Moreover little attention has
been given to the significance of the unpaid labour applied to the built
environment in the creation of gentrified dwellings.

Thirdly, and most importantly, there are ambiguities as to the role
of human agency in the process itself. In early work on gentrification
it is possible to identify two major, and apparently conflicting, forms
of explanation. The first, associated with Marxist urban theory, examined factors relating to the supply or production of housing. Attention
focused on abstract, structural forces underpinning a twofold shift in
patterns of investment: there has been a shift between investment
sectors, away from manufacturing industry towards the built environment and to the service sector (Harvey 1978); and there has been a
spatial movement of investment away from suburban towards inner-
city (and to rural) locations (Smith 1982). Within this supply-side or
structuralist perspective, an attempt is made to unravel the logic of
these trends and to understand them in terms of "contradictions" of
capitalism. The second major approach to gentrification has closer
links with positivism and behavioural approaches. It focuses on
changes in demand for, or consumption of, different residential forms
and locations, which are analyzed through demographic trends and
individual preferences and decisions. From this perspective, gentrification is understood in terms of the housing and lifestyle choices of a
generation born in the postwar baby boom.

The weaknesses of both approaches are now well documented (see
Rose 1984; Smith and Williams 1986; Munt 1987). Briefly the first,
with its emphasis on the inexorable logic of capital, tends to become
teleological and to reduce human actors to the status of "puppets"
controlled by "capitalist puppeteers" (Eyles 1987). Conversely the
second underestimates the role of factors constraining human choices
and tends to reduce the explanation of social phenomena to the
behaviour of individuals. In both approaches these respective weaknesses result in an inadequate analysis of the relationship between
gentrification and changes in the organization of work. The structuralist approach ignores the way in which such changes are negotiated

and interpreted by conscious human actors, and implicitly disregards unpaid work by emphasizing forces underlying changing patterns of employment. The behaviouralist approach reduces work to an individual attribute, namely occupational type, largely ignoring both the limitations of occupational classifications and forms of work other than employment.

In response to these criticisms a number of writers have developed conceptualizations that treat gentrification as a social process through which changes in the organization of work, within both the formal and domestic economies, are expressed in the urban environment. For example Ley (1980) emphasizes the role of a new professional elite as agents expressing an ideology of "livability" in opposition to an ideology of economic growth. From a neo-Marxist perspective, Williams (1986a) examines gentrification as "class constitution," that is, as a process through which changes in the structure of capitalism and of class relations are realized and expressed by changes in the class identity of human agents. Despite differences in detail, these studies illustrate the principle of combining constraints and choices, or social structures and human agency, within explanations of gentrification. The idea that gentrification expresses changes in the relationship between home and work is also implicit: new kinds of paid work generate distinctive social groups that express their identity, at least in part, by creating distinctive home environments.

Related to the emergence of such approaches is a more explicit reappraisal of the relationship between the built environment, in the form of owner-occupied housing, and social divisions. On the one hand, increasing emphasis has been given to the importance of housing as an expression of social identity (Saunders 1984; Jager 1986); thus gentrification can be viewed as expressing the emergence of a new social group and, therefore, as related to both deep-seated structural change in society and to the projection of particular images connoting difference and diversity (Smith 1987). On the other hand, housing has been recognized as an increasingly important factor underlying social divisions. In particular the argument advanced by Dunleavy (1979) that a new social cleavage can be identified in British society – based on access to private as opposed to public means of consumption and irreducible to class divisions – has become influential (see for example Duke and Edgell 1984; Saunders 1984, 1986). This approach has not escaped criticism; from differing perspectives Harrison (1986), Klausner (1986), Pratt (1986), Preteceille (1986), Smith (1987), Bondi and Peake (1988), and Mackenzie (1988) all stress the importance of understanding the interplay between consumption and production. Nevertheless it is widely accepted that occupational status is declining in significance as a factor underlying social divisions, while housing and residence are, conversely, becoming more important (Williams 1986b; Forrest 1987).

Given the powerful and widespread association between women and the home, such changes are bound to raise questions pertaining to gender divisions (see Loyd 1981; Bondi and Peake 1988). Consequently it is not surprising that (with the exception of the most abstract, structuralist approaches), references to women are common in explanations of gentrification. These claims are now examined.

FROM WOMEN AND GENTRIFICATION TO THE "PRODUCTION" OF FEMALE GENTRIFIERS

Within the literature on gentrification references to the role of women first emerged in the context of studies concerned with the demand for gentrified housing. Propositions arising from these writings can be divided according to those relating to women's work in the formal economy and those relating to in the domestic economy.

Changes in the formal economy. It has been suggested that gentrification is being stimulated by the increased participation of women (especially married women) in the labour force. Of particular significance is the increasing success of middle-class women in obtaining well-paid career jobs. In other words women drawn from particular social groups are swelling the ranks of the "yuppies" (Wekerle 1984). This creates relatively more affluent households, whether consisting of single women or of married or cohabiting couples, and it generates demand for increasingly expensive private housing. A different economic argument is advanced by Rose (1984), who suggests that gentrification is also being stimulated by first-time buyers on moderate incomes and employed in city centres, typically in service sector industries. This group is suggested to include lone women and households headed by women following of marital breakdown. This line of argument has been supplemented by a suggestion that gentrification has been fuelled by women's preference for central residential locations. This preference is sometimes claimed of all women and sometimes only of the well-off and childless (Holcomb 1984; Wekerle 1984). Implicit in this assumption is an hypothesis to the effect that the increasing participation of women in the labour market is giving women greater economic leverage and, therefore, greater influence in decision making within households.

Changes in the domestic economy. Changes in the position of women in the labour market are closely related to changes in the position of women in the family and the domestic economy. Accounts of gentrification emphasizing the latter take two basic forms. First, it has been suggested that gentrification is the result of an increase in the absolute number of households in many Western societies, which is accounted for, in a large part, by an increase in the number of women living alone. Important factors here include the increasing number and independence of the elderly, among whom women predominate; the increased rate of divorce; and the rising average age of marriage. Thus there are increasing numbers of households consisting of women who are elderly and widowed, divorced, or young and not (yet) married. The last category is itself enlarged by the arrival at adulthood (or, more precisely, at house-purchasing age) of the postwar baby-boom generation. These women do not undertake domestic work for adult men in their own homes, and women in two of the three categories do not have immediate responsibility for children. It should be noted that the people identified in this argument may not be gentrifiers themselves but do contribute to total demand for housing units.

The second economic explanation of gentrification emphasizes the changing character of family units, in particular the postponement of childbearing, decreased completed family size, and closer spacing of children (Martin and Roberts 1984). Combined with changes in women's employment, this reduction in the burden of child care creates more small, affluent households (sometimes known as "dinkys": dual-income-no-kids-yet) capable of paying high prices for sought-after housing. Moreover these households benefit more from reduced commuting costs associated with inner-urban residential locations than do those with only one adult working in the city centre. In some areas a similar demand is generated by dual-career gay households (Markusen 1981; Lauria and Knopp 1985).

There is some empirical evidence to support each of these propositions, but it remains fragmentary and attempts to explore the interconnections and interplay among the various elements of women's role remain limited. Multiple regression models such as that offered by Ley (1986) fail to illuminate the substantive links among the phenomena concerned. Instead, like all positivist-based approaches, such analyses retain a mechanical view of work, which is equated with occupational type, and of the categories of women and men. This inhibits examination of the social processes producing both gentrifiers and gendered individuals. Clearly satisfactory investigation of these propositions must be rooted in a broader analysis of the changing position of women in society and the changing nature of work.

Analyses of gentrification that stress the explanatory importance of the "production and reproduction" or "constitution" of gentrifiers are the most promising for such an endeavour because of the attempt to link the construction of particular social identities with underlying structural forces. In terms of women and gentrification Smith (1987, 164) suggests a potential point of departure for study. Accepting that the rise of the dual-income households provides a logic for inner-urban residential location, he stresses the need to explain why the gradual quantitative increase in the proportion of women working and women in higher-income brackets translates into a substantial spatial change of domicile. After all, married women were in the official work force before World War II, albeit in smaller numbers; some of these were in relatively well-paying professional positions, yet no gentrification process seems to have blunted the suburban flight of the time. How could such a comparatively quick spatial reversal be explained by more gradual social changes alone?

Smith proceeds to hint at the significance of the contemporary feminist movement. But he stops short of considering gender identity as an important aspect of the constitution of gentrifiers, arguing instead that while changes in the position of women as paid workers have made gentrification possible, the critical factor is that of the cultural differentiation of classes: "[G]entrification is a redifferentiation of the cultural, social, and economic landscape, and, to that extent, one can see in the very patterns of consumption clear attempts at social differentiation. Thus gentrification and the mode of consumption it engenders are an integral part of class constitution ... they are part of the means employed by new middle-class individuals to distinguish themselves from the stuffed-shirt bougeoisie above and the working class below [in their different but equally homogeneous suburbs]" (1987, 168). This shifts the argument back to the familiar territory of class. Despite his apparent interest in the role of women in gentrification, Smith underplays it in two ways. First, he largely ignores the domestic economy and therefore fails to capture the way in which women's participation in waged labour influences their lives. Secondly, he ignores the impact of changes in women's activities on men and therefore fails to examine gender relations. In short he treats gender inequalities as derivative of capitalist social relations.

Rose (1989) offers an advance on Smith's conceptualization in her argument that gender practices, in the form of women's employment, may be constitutive of gentrification. However, although she recognizes that gender divisions are not reducible to class relations, she limits consideration of gender practices to patterns of employment, and she, too, does not consider the masculine contribution to gender divisions. As these criticisms suggest, a more successful approach to the role of women in gentrification requires the adoption of a concept of gender that is relational and that can contribute to discussions of the constitution of gentrifiers. The next section attempts to sketch out such a conceptualization.

GENDER AS A SOCIAL PRACTICE

I have argued thus far that the most successful approaches to analysis of gentrification seek to transcend the dichotomy of agency and structure. Here a parallel argument will be advanced in relation to gender, drawing particularly on Connell's (1987) critique and reformulation. According to Connell current frameworks for the analysis of social divisions based on gender can be divided into two main groups: there are extrinsic theories, which explain inequalities between women and men in terms of factors or structures external to gender identity, such as those underpinning gender divisions of labour (human behaviour, including differences in gender identity, is explained in terms of these structures); and there are intrinsic theories, which assume that the constitution of gender is itself the basis for gender inequalities. Within this latter approach, causality is attributed to gender identity per se and by implication to the actions and experiences of individuals.

Extrinsic theories focus on broad, systemic patterns of female subordination and are best exemplified by various versions of Marxism and socialist feminism. The former treat gender inequality as derivative of capitalism and, as illustrated above, are implicit in some analyses of gentrification. Socialist feminists have largely rejected this position and have developed interpretations of women's subordination in terms of the interrelations and interactions between capitalism and patriarchy. Several such formulations have in common their refusal to subsume the oppression of women within a class analysis, positing instead a gender structure equally fundamental to and at least partially independent of class structure. Walby (1986) distinguishes between "dual systems" theories that treat capitalism and patriarchy as operating in different spheres and those that treat both as ubiquitous. The former category includes theorists who view patriarchy as an ideological or cultural structure as opposed to the economic structure of capitalism (e.g., Mitchell 1974) and theorists who differentiate between production and reproduction as the realms of capitalism and patriarchy respectively (e.g., Foord and Gregson 1986); in both cases gender inequalities in the workplace are assumed to originate elsewhere. Other variants of dual systems theories interpret capitalist and patriarchal social relations as ubiquitous but independent (e.g., Hartmann 1979; Walby 1986). This allows for greater flexibility in interpreting the sources of gender divisions within the formal and the domestic economy but tends to result in a vague and often confused notion of patriarchy (Beechey 1976; Foord and Gregson 1986).

The dual systems approach was crucial to the defence by socialist feminists of an autonomous women's movement and to challenge the sexism of left-wing men (see Rowbotham, Segal, and Wainwright 1979). By treating gender as a phenomenon parallel or analogous to class, however, these accounts suffered the same methodological weaknesses as did the Marxist formulations on which they drew. Functionalism was sometimes apparent (albeit sometimes defended as a necessary first step in the development of a new theory, as by Barrett [1980]). Elsewhere the attempt to identify the independent origins of patriarchy floundered on an implicit appeal to biological imperatives (e.g., Foord and Gregson 1986). More generally the exclusive concern with systems or structures external to individuals begged questions concerning gender identity, consciousness, and social change.

Connell (1987) identifies two main types of intrinsic theories: role theory, and what he terms "categorical" theories, which include radical feminist and some socialist-feminist formulations that emphasize "identity politics." Role theory analyzes social life as a drama in which people occupy and internalize positions to which particular behaviours are allocated. The performance of these roles is maintained through a system of expectations and sanctions experienced and enforced by individual human agents. Sex is one of the criteria used to allocate roles, and the various social behaviours associated with each sex generate distinctive "gender roles" (Oakley 1972). Girls and boys are socialized into these roles by "significant others" who occupy positions within such institutions as the family, the education system, the media, and so on. Thus children acquire gender identities and, according to role theory, broad patterns of gender differentiation and gender inequality are the product of these gender identities.

Role theory is attractive in its attempt to transcend the structure/agency dichotomy. The concept of gender roles has also served an important function in feminist analyses, where it continues to be very widely used to demonstrate the social basis of what are often perceived as natural differences in the work that women and men do. However the approach also suffers certain weaknesses. It tends towards functionalism in that role differences based on sex are often implicitly treated as socially necessary and, indeed, desirable, and it is overdetermined in that it implies gender divisions are both structurally determined and actively created. In addition, and following from this overdetermination, the theory lacks a dynamic and therefore cannot explain social change. Lastly, role theory adopts a very limited view of power: power is evident only in the concepts of peer pressure and the reference group; systemic forms of power based on gender or class are not admitted.

By contrast categorical theories accord a central role to the power of particular groups. In radical feminism women and men are identified as the basic groups whose interests are opposed and structured by a systemic power imbalance. This power relation reinforces and repro-duces the oppression of women, and its key mechanisms are intrinsic to gender: gender identity is cited as an explanation for female subor-dination. Thus Brownmiller (1976, 15) describes rape as a "conscious process of intimidation by which all men keep all women in a state of fear." Accordingly, in the radical feminist account, differences between the categories "women" and "men" are much greater than differences within these categories, and therefore all women have interests in common that override interests that diverge. This view has been widely challenged. One alternative developed particularly by socialist femi-nists has been to introduce cross-classifications based on class and race (see Young 1981; Vogel 1983). This merely serves, however, to high-light the fundamental difficulty of categorical theories: neat boxes limit the analysis to static categories, the constitution or transformation of which becomes all the more elusive. This difficulty is well illustrated by Connell (1987, 60) in connection with sexual orientation.

Heterosexism and homophobia must be regarded as one of the key attitudinal patterns in gender relations. It is very difficult to adequately address heterosexual dominance while using a categorical model of gender. One might set up a new cross-classification between "homo/ hetero" and "female/male," but there is no obvious reason in categor-ical analysis why this division should matter in the first place. Cross-classification confuses the issue, since that operation logically equates women's homosexuality with men's homosexuality. There is every reason to think there are important differences between them, not only in the forms of their expression but in the ways they are initially constituted. As the chequered experience of lesbians in the gay libera-tion movement shows, solidarity of women and men against oppres-sion from straight society cannot be taken for granted. That is to say, logical cross-classification does not correspond in any simple way to the constitution of a social interest. Moreover the categorical approach underlies much of the fragmentation associated with "identity poli-tics," in which gender, class, sexual orientation, race, ethnicity, and other elements of identity form the basis for group organization (Adams 1989).

To conceptualize gender (or elements of it) as a category thus limits rather than enhances understandings of both its constitution and its role in other social phenomena, such as gentrification. Concern with the constitution of gender implies that the boundary between what it means to be a woman and what it means to be a man (femininity and masculinity) is changing and is problematical. Further, to investigate the constitution of gender it is necessary to combine consideration of structure, addressed in extrinsic theories, and identity, which intrinsic theories highlight. Within such an approach, gender can be conceptualized in terms of an interweaving of, and recursive relationships between, personal life and conscious human action, on the one hand, and social structures or constraints, on the other. As Connell (1987) observes, this conceptualization is illustrated evocatively in many feminist novels, but formal theorizing is very limited. A model is provided by theories of practice, such as those advanced by Giddens (1979, 1984) and Bourdieu (1977). Although detailed discussion of how these might be applied to the analysis of gender is beyond the scope of this study, the following discussion serves to illustrate the approach.

First, femininity and masculinity are not merely individual attributes influencing or explaining particular choices or actions; they are also social phenomena structuring all aspects of the lives of human subjects. In other words gender exists both as an aspect of individual, subjective identity and as an external, social construct that constrains the behaviour of women and men – hence the need to combine intrinsic and extrinsic approaches. Secondly, these internal and external definitions of gender are neither unchanging nor in perfect harmony. At its simplest, this situation creates scope for role reversal, which occurs when an individual assumes a role normally associated with the opposite sex. But the consequences can be understood more widely in relation to the argument advanced by Bowlby, Foord, and McDowell (1986) that geographers have begun to consider the significance of gender roles in their research but have paid inadequate attention to gender relations. The distinction is explained as follows: "By gender roles we mean a relatively static set of assumptions about the supposed characteristics of women and men, rather than an active social process of gender relations through which male power over women is established and maintained" (Bowlby, Foord, and McDowell 1986, 328). Thus relations between women and men are continually being redefined and renegotiated. The tension and interplay between internalized and structural aspects of gender can be viewed as the driving force for changes in gender relations (see Figure 1).

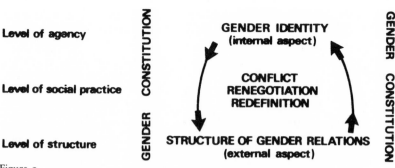

Figure 1
Structure and Agency in the Constitution of Gender

CONCLUSION: GENTRIFICATION, WORK, AND GENDER IDENTITY

Existing propositions about women and gentrification refer, implicitly, to gender roles and do not consider the social practices through which gender is constituted. Further, changes in gender roles (such as the entry of married women into the labour force) are most often assumed to derive from changing class relations rather than changing gender relations. Consequently the femininity of domestic work, and of particular occupations, is treated as a static category, apparently external to the process of gentrification. A more illuminating approach would treat gender as a fluid and dynamic phenomenon involving changing identities and power relations as well as changes in the roles of women and men. In terms of gentrification this suggests that an understanding of the significance of gender must be sensitive to the ways in which gender is constituted within social practices, including paid employment and unpaid domestic work. For example, insofar as the position of women in the labour market is significant, it is important to examine how changes in the relative economic positions of women and men are translated into changes in decisions about home life. Such decisions include those relating to divisions of labour within the household, the choice between domestic labour and services bought in the formal economy, and house purchase itself.

In addition to changes within the home precipitated by the success of middle-class women in the labour market, other changes in the domestic sphere merit investigation. In particular the affluence of small professional households is providing a focus for much consumer production and advertising. A great deal of this marketing centres on the cache of particular lifestyles, packaged and delivered through architectural styles, interior design, and consumer durables (Mills 1988). One of the subtleties of this marketing is the message that the consumer actively creates his or her own distinctive environment. This implies that domestic work and homemaking are being transformed in terms of meaning as well as content, which in turn may generate new understandings of gender identity in the form of ideas about masculinity and femininity. Existing evidence suggests that women generally play an important part in decisions about consumption, especially concerning the interior of homes and possibly in the choice of a house to buy or rent (Loyd 1981). This reflects the role of women as domestic workers and homemakers. It is therefore necessary to examine the relationship between decisions about housing that result in gentrification, attitudes to homemaking, and understandings of gender. More specifically, insofar as women's preferences are reflected in decisions resulting in gentrification, an important question remains: does the expression and realization of these preferences reinforce, reduce, modify, or leave unaltered existing gender divisions within the home, or in any way encapsulate new conceptions of masculinity and femininity?

The exploration of such issues entails examining the internal structure and operation of the household as a decision-making unit containing gendered individuals whose daily lives contribute to the reproduction, reconstitution, and redefinition of gender identities and gender relations. Such an approach stands in stark contrast to the geographical convention of treating the household as an undifferentiated object (Tivers 1978). In conclusion the argument advanced here is for a shift in focus: rather than formulating questions in terms of the role of women in gentrification, it may be more fruitful to consider how the social constructs of femininity and masculinity are expressed and renegotiated in the built environment. This essay has attempted to indicate how this might be achieved by drawing a parallel between the constitution of gentrifiers and the constitution of gendered individuals. Approaching both issues in terms of social practices in which individual identities and external constraints are mutually constructed allows the structure/agency dichotomy to be transcended. In this way it may be possible to understand gentrification as one of several spatial expressions of the changing character of work in women's lives.

REFERENCES

Adams, M.L. 1989. "There's no place like home: on the place of identity in feminist politics." *Feminist Review* 31: 22–33.
Barrett, M. 1980. *Women's Oppression Today.* London: Verso.

Beauregard, R.A. 1986. "The chaos and complexity of gentrification." In *Gentrification of the City*, N. Smith and P. Williams, eds, 35–55. Boston: Allen and Unwin.

Beechey, V. 1976. "On patriarchy." *Feminist Review* 3: 66–82.

– 1986. "Women's employment in contemporary Britain." In *Women in Britain Today*, V. Beechey and E. Whitelegg, eds, 77–131. Milton Keynes: Open University Press.

Bondi, L. and L. Peake. 1988. "City and society: a feminist reformulation of urban politics." In *Women in Cities*, J. Little, L. Peake, and P. Richardson, eds, 21–40. London: Macmillan.

Bourdieu, P. 1977. *Outline of a Theory of Practice*. Cambridge: Cambridge University Press.

Bowlby, S.R., J. Foord, and L. McDowell. 1986. "The place of gender relations in locality studies." *Area* 18: 327–31.

Brownmiller, S. 1976. *Against Our Will: Men, Women and Rape*. Harmondsworth: Penguin.

Connell, R.W. 1987. *Gender and Power*. Cambridge: Polity Press.

Davidoff, L., J. L'Esperance, and H. Newby. 1976. "Landscape with figures: home and community in English society." In *The Rights and Wrongs of Women*, J. Mitchell and A. Oakley, eds, 139–75. Harmondsworth: Penguin.

Duke, V. and S. Edgell. 1984. "Public expenditure cuts in Britain and consumption sectoral cleavages." *International Journal of Urban and Regional Research* 8: 177–201.

Dunleavy, P. 1979. "The urban bases of political alignment." *British Journal of Political Science* 9: 409–43.

Eyles, J. 1987. Book review. *Transactions of the Institute of British Geographers*, n.s., 12: 116–20.

Foord, J. and N. Gregson. 1986. "Patriarchy: towards a reconceptualisation." *Antipode* 18: 186–211.

Forrest, R. 1987. "Spatial mobility, tenure mobility and emerging social divisions in the U.K. housing market." *Environment and Planning A* 19: 1611–30.

Freeman, J. 1981. "Women and urban policy." In *Women and the American City*, C.R. Stimpson, E. Dixler, M.J. Nelson, and K.B. Yatrakis, eds, 1–19. Chicago: University of Chicago Press.

Gershuny, J. 1983. *Social Innovation and the Division of Labour*. Oxford: Oxford University Press.

Giddens, A. 1979. *Central Problems in Social Theory*. London: Macmillan.

– 1984. *The Constitution of Society. An Outline of the Theory of Structuration*. Cambridge: Polity Press.

Harrison, M.L. 1986. "Consumption and urban theory: an alternative approach based on the social division of welfare." *International Journal of Urban and Regional Research* 10: 232–42.

Hartmann, H.I. 1979. "The unhappy marriage of Marxism and feminism: towards a more progressive union." *Capital and Class* 8: 1–33.

Harvey, D. 1978. "The urban process under capitalism: a framework for analysis." *International Journal of Urban and Regional Research* 2: 100–31.

Holcomb, B. 1984. "Women in the rebuilt urban environment: the United States experience." *Built Environment* 10. 1: 18–24.

Jager, M. 1986. "Class definition and the esthetics of gentrification: Victoriana in Melbourne." In *Gentrification of the City*, N. Smith and P. Williams, eds, 78–91. Boston: Allen and Unwin.

Klausner, D. 1986. "Beyond separate spheres: linking production with social reproduction and consumption." *Environment and Planning D: Society and Space* 4: 29–40.

Lauria, M. and L. Knopp. 1985. "Toward an analysis of the role of gay communities in the urban renaissance." *Urban Geography* 6: 152–69.

Lewis, J. 1984. "The role of female employment in the industrial restructuring and regional development of the United Kingdom." *Antipode* 16. 3: 47–59.

Lewis, J. and J. Foord. 1984. "New towns and new gender relations in old industrial regions: women's employment in Peterlee and East Kilbride." *Built Environment* 10. 1: 42–52.

Ley, D. 1980. "Liberal ideology and the post-industrial city." *Annals of the Association of American Geographers* 70: 238–58.

– 1981. "Inner-city revitalization in Canada: a Vancouver case study." *Canadian Geographer* 25: 124–48.

– 1986. "Alternative explanations for inner-city gentrification: a Canadian assessment." *Annals of the Association of American Geographers* 76: 521–35.

Loyd, B. 1981. "Women, home and status." In *Housing and Identity: Cross Cultural Perspectives*, J.S. Duncan, ed, 181–97. London: Croom Helm.

McDowell, L. 1983. "Towards an understanding of the gender division of urban space." *Environment and Planning D: Society and Space* 1: 59–72.

Mackenzie, S. 1988. "Building women, building cities: toward gender sensitive theory in the environmental disciplines." In *Life Spaces: Gender, Household, Employment*, C. Andrew and B.M. Milroy, eds, 13–30. Vancouver: University of British Columbia Press.

Mackenzie, S. and D. Rose. 1983. "Industrial change, the domestic economy and home life." In *Redundant Spaces in Cities and Regions*, J. Anderson, S. Duncan, and R. Hudson, eds, 155–200. London: Academic Press.

Markusen, A. 1981. "City spatial structure, women's household, and national urban policy." In *Women and the American City*, C.R. Stimpson, E. Dixler, M.J. Nelson, and K.B. Yatrakis, eds, 20–41. Chicago: University of Chicago Press.

Martin, J. and C. Roberts. 1984. *Women and Employment: A Lifetime Perspective*. London: HMSO.

Massey, D. 1983. "Industrial restructuring as class restructuring: production decentralization and local uniqueness." *Regional Studies* 17: 73–89.

Mills, C.A. 1988. "'Life on the upslope': the postmodern landscape of gentrification." *Environment and Planning D: Society and Space* 6: 169–89.

Mitchell, J. 1974. *Psychoanalysis and Feminism*. Harmondsworth: Penguin.

Molyneux, M. 1979. "Beyond the domestic labour debate." *New Left Review* 116: 3–27.

Munt, I. 1987. "Economic restructuring, culture, and gentrification: a case study in Battersea, London." *Environment and Planning A* 19: 1175–97.

Oakley, A. 1972. *Sex, Gender and Society*. London: Temple Smith.

Palen, J. and B. London. 1984. *Gentrification, Displacement and Neighborhood Revitalization*. Albany, NY: State University of New York Press.

Pratt, G. 1986. "Against reductionism: the relations of consumption as a mode of social structuration." *International Journal of Urban and Regional Research* 10: 377–400.

Preteceille, E. 1986. "Collective consumption, urban segregation and social classes." *Environment and Planning D: Society and Space* 4: 145–54.

Rose, D. 1984. "Rethinking gentrification: beyond the uneven development of Marxist urban theory." *Environment and Planning D: Society and Space* 1: 47–74.

– 1989. "A feminist perspective of employnment restructuring and gentrification: the case of Montreal." In *The Power of Geography*, J. Wolch and M. Dear, eds, 118–38. Boston: Unwin Hyman.

Rowbotham, S., L. Segal, and H. Wainwright. 1979. *Beyond the Fragments: Feminism and the Making of Socialism*. London: Islington Community Press.

Saegert, S. 1981. "Masculine cities and feminine suburbs: polarized ideas and contradictory realities." In *Women and the American City*, C.R. Stimpston, E. Drixler, M.J. Nelson, and K.B. Yatrakis, eds, 93–108. Chicago: University of Chicago Press.

Saunders, P. 1984. "Beyond housing classes: the sociological significance of private property rights and means of consumption." *International Journal of Urban and Regional Research* 8: 202–27.

– 1986. "Comment on Dunleavy and Preteceille." *Environment and Planning D: Society and Space* 4: 155–64.

Smith, N. 1982. "Gentrification and uneven development." *Economic Geography* 58: 139–55.

– 1987. "Of yuppies and housing: gentrification, social restructuring, and the urban dream." *Environment and Planning D: Society and Space* 5: 151–72.

Smith, N. and P. Williams, eds. 1986. *Gentrification of the City*. Boston: Allen and Unwin.

Thrift, N. 1987. "The geography of late twentieth century class formation." In *Class and Space: The Making of Urban Society*, N. Thrift and P. Williams, eds, 207–53. London: Routledge and Kegan Paul.

Tivers, J. 1978. "How the other half lives: the geographical study of women." *Area* 10: 302–6.

– 1985. *Women Attached: The Daily Lives of Women with Young Children*. Beckenham: Croom Helm.

Vogel, L. 1983. *Marxism and the Oppression of Women: Toward a Unitary Theory*. London: Pluto Press.

Walby, S. 1986. *Patriarchy at Work*. Cambridge: Polity Press.

Wekerle, G.R. 1984. "A woman's place is in the city." *Antipode* 16. 3: 11–19.

Williams, P. 1984. "Gentrification in Britain and Europe." In *Gentrification, Displacement and Neighborhood Revitalization*, J. Palen and B. London, eds, 205–34. Albany, NY: State University of New York Press.

– 1986a. "Class constitution through spatial reconstruction? a re-evaluation of gentrification in Australia, Britain and the United States." In *Gentrification of the City*, N. Smith and P. Williams, eds, 56–77. Boston: Allen and Unwin.

– 1986b. "Social relations, residential segregation and the home." In *Politics, Geography and Social Stratification*, K. Hoggart and E. Kofman, eds, 247–73. Beckenham: Croom Helm.

Young, I. 1981. "Beyond the unhappy marriage: a critique of the dual systems theory." In *Women and Revolution*, L. Sargent, ed, 43–69. London: Pluto Press.

Contributors

HAL BENENSON is an associate professor of sociology at McGill University. He is the author of *Dependent Wives and "Family Wage": Men in Twentieth-Century Britain*, forthcoming with Polity Press.

LIZ BONDI is a lecturer in geography at the University of Edinburgh. She researches and publishes in the field of feminist geography and coedits the journal *Gender, Place and Culture*.

BETTINA BRADBURY teaches history and women's studies at Glendon College, York University. She has published many articles on the history of women and the family in the period of industrialization and is the author of *Working Families: Age, Gender and Daily Survival in Industrializing Montreal* (1993), as well as editor of *Canadian Family History: Selected Readings* (1992).

SUSAN CHRISTOPHERSON is an associate professor in the City and Regional Planning Department at Cornell University. She has published widely in the areas of labour markets, women's employment, and media studies and is completing a book on how financial-market rules in the United States affect firm investment behaviour. She is actively engaged in international policy research for the Organization for Economic Cooperation and Development and for the United Nations Conference on Trade and Development. She is currently completing research for a study of the implications of media globalization and trade policy for China for UNCTAD.

SYLVIA GOLD is a faculty member of the Canadian Centre for Management Development in Ottawa, and is the coordinator of the centre's management trainee program. From 1985–89 she was the president of the Canadian Advisory Council on the Status of Women. During the years 1990–92 she was director of research and evaluation for the Royal Commission on New Reproductive Technologies. She has an MA from McGill University and has completed course work for an education degree at OISE.

ALISON KAYE received her PhD From Queen Mary Westfield College, and has taught at Birkbeck College, University of London. She is currently working on a project entitled "Bilingual Women: Educational and Training Needs," a study of two groups of women, South Asian and Somali, living in west London. The study is based at the Centre for Extramural Studies, London University.

AUDREY KOBAYASHI is director of the Institute of Women's Studies and professor of geography at Queen's University. She is active in community work and has published widely on Japanese Canadians and on social issues of racism and gender. She is coeditor, with Suzanne Mackenzie, of *Re-making Human Geography* and is currently completing a project, with Peter Jackson, on racism in Britain and Canada and another on Japanese immigrants in Canada.

JOY PARR is Farley Professor of History at Simon Fraser University. Her most recent book is *Gender of Breadwinners: Women, Men and Change in Two Ontario Towns* (1990). She is currently writing a comparative economic history of consumption and consumer goods in postwar Canada and Sweden. In 1992 Joy was elected a fellow of the Royal Society of Canada.

LINDA PEAKE is an associate professor in the Social Science Division at York University. She has published widely in the fields of gender development and theory and is currently working on a project with low-income women in Guyana, investigating the effects of structural adjustment policies on their daily lives.

KATIE PICKLES is a graduate student in geography at McGill University. Her doctoral thesis is an historical/cultural geography of imperialism and citizenship through a case study of the Imperial Order Daughters of the Empire (IODE).

DAMARIS ROSE is an urban and social geographer. She is an associate professor at the Institut national de la recherche scientifique – Urbanisation in Montreal. Her research interests include gentrification and socioeconomic polarization in inner cities, the social impacts of industrial restructuring, and gender and urban theory. She has published numerous articles in these fields and is coauthor of a forthcoming book entitled *Montréal: The Quest for a Metropolis*.

PAUL VILLENEUVE teaches human geography at Université Laval, where he also heads the Centre de recherche en aménagement et développement. He has published widely in the area of Canadian social geography.

Index

abortion, 100; legalization of, xv
Afro-Caribbean women, 113
agriculture, 171
Althusser, Louis, and Marxism, 5–6
androcentric, worldviews, 5
androcentrism, and social change, 11
antinuclear lobby, 14
Antipode Debate, xxx
apprentices, 33
Arizona, households in nonmetropolitan areas of, xxxiii; nonmetropolitan development in, 167–73
Asian women, 16–17
automation, and reskilling, 134

bakeries, commercial, 41
Bangladeshis, 112–27 passim
behavioural theories, 186
benefits, 171
Bengali, 126 n.2
biological imperatives, 192

biologism, 14
biology, and naturalism, xix
Bisbee, Arizona, 172–4
Black households, in inner-cities, 185
Black women, 16–17; in the garment industry, 121
bluestocking, 108
boardinghouse keepers, 85
bookbinding, 33
Brantford, Ontario, 76, 80, 88–9
bread, purchase of, 41
breadwinner, male wage, 83; men as, 183
Britain, xxxii, xxxv, xxxv n.4, caregivers in, 17; change in women's employment in, 18; compared to Japan, 46; discrimination in labour market, 113; dismantling of marriage bar in, xviii; garment industry, 121; industrialization, xvi; industrial restructuring in, 184; political repre-

sentation in, 99; postwar, 14; and suffrage, xv; waged work in, 14; women's movement, xvi; women's work in, 3
British Columbia, immigration and economy, 58
Buddhism, 60
built environment: new forms of, xxiv; and rebuilding, xxviii; refashioning of, xvi; renegotiation of, 196
business: management, 135; service sector, 173
buttonhole machine, 35

Canada, xxxv; celebration of International Women's Day, 101; garment factories, xxv; and women's work, xxxiii
Canadian Advisory Council on the Status of Women, 105
Canadian Jobs Strategy, 105